BEEKEEPING STUDY NOTES
for the
BASIC AND INTERMEDIATE EXAMINATIONS OF THE FIBKA

Prepared by:

J. D. Yates B.Sc (Hons), C. Eng., FIEE

and

B. D. Yates SRN. SCM.

Northern Bee Books

Beekeeping Study Notes for the Basic and Intermediate
Examinations of the FIBKA
© B. D. Yates

Reprinted 2015

Northern Bee Books
Scout Bottom Farm
Mytholmroyd
Hebden Bridge
HX7 5JS (UK)

ISBN 978-1-908904-72-0

D&P Design and Print
Worcestershire

Printed by Lightning Source, UK

BEEKEEPING
STUDY NOTES
for the
BASIC AND INTERMEDIATE
EXAMINATIONS OF THE FIBKA

Prepared by:

J. D. Yates B.Sc (Hons), C. ENG., FIEE

and

B. D. Yates SRN. SCM.

Northern Bee Books

First published 1990

Second edition 1991

Notes to the Second Edition

As a result of many helpful comments received from various readers of the first edition, this edition has been made physically smaller and has been type-set directly from the original computer discs thereby eliminating many minor errors which appeared originally. A major comment has been that all three essential books, which we recommended, are out of print. We believe that there are no substitutes available for Dade and Dadant, and second hand copies will have to be sought and purchased. Hooper's book may be substituted with 'Beekeeping - A Seasonal Guide' by Ron Brown. Both books are very similar and deal with general beekeeping in the UK although Brown's book tends to be orientated to beekeeping in Devon and has more practical beekeeping suggestions.

JD & BD Yates
Newton Ferrers
Devon
1991

FOREWORD

Not least among the services and endeavours of the BBKA is the function of the Examination Board. Systematic study of our fascinating craft yields the satisfying reward of a better understanding of bees, both in regards to practical beekeeping and the scientific aspects of bees and the craft. The Board establishes a syllabus for each of the three examinations leading to the award of the Master Beekeeper's Certificate, and to assist students the Board also provides the facility of a Correspondence Course. Guidance is given in regard to suitable books, and each student is allocated to a tutor.

Along with other experienced beekeepers I act as a tutor for the Course. In this capacity I find that often students are delayed and inconvenienced by not being able to borrow the recommended books, very often, it seems to me, because Association branches simply do not make a practice of maintaining a good library for the general use of members. John and Dawn Yates fulfil a much needed requirement in this connection by providing a detailed and thorough narrative systematically covering every aspect of the Intermediate Examination syllabus. Their beekeeping experience and expertise, passed on through this volume, not only caters for methodical Intermediate study, but is also a good reference for Basic students.

Senior students too can usefully have a copy at hand. Not all beekeepers have an interest in the BBKA examinations, but they also can certainly reap a good deal of benefit from this publication, which to my mind is long overdue.

Frank Alston.
Newlands, Budleigh Salterton.

CONTENTS

3.6   An elementary description of the process of fermentation in honey.
3.7   An account of the use of nectar, honey and water by the honeybee colony.
3.8   An account of how pollen is collected, carried and stored.
3.9   An account of the importance of pollen in the nutrition of the honeybee.
3.10   An account of the collection and use of propolis by the honeybee.

4.0 Flora and pollination.
4.1   An account of the main nectar and pollen producing plants of the British Isles and their flowering periods.
4.2   An elementary account of the process of pollination and fertilisation of a flowering plant.
4.3   An account of the honeybee as a pollinating insect and of its usefulness to farmers and growers.

5.0 Disease and poisoning.
5.1   A detailed account of the field diagnosis of American Foul Brood (AFB) and European Foul Brood (EFB) and a tabulation of the differences between the signs of these two diseases.
5.2   An account of the ways in which foul brood infections can spread from colony to colony.
5.3   A detailed account of action necessary to take when AFB or EFB is found, including treatments and sterilisation of equipment.
5.4   A detailed account of the method of detection of Varroa jacobsoni and its differentiation from Braula coeca.
5.5   An account of the major legal requirements relating to Foul Brood, Varroasis and the importation of bees applicable to England and Wales.
5.6   A detailed account of the signs of Chalk Brood, Sac Brood, Bald Brood, Varroasis, Addled Brood and Stone Brood.
5.7   A detailed account of the signs of and the recommended treatment for adult bee diseases.
5.8   An elementary knowledge of Chronic Bee Paralysis Virus (Syndromes type 1 and 2) and Black Queen Cell Virus and their association with other diseases.
5.9   An account of Chilled Brood and its possible causes.
5.10   An account of colony starvation and possible remedial actions.
5.11   An account of the poisoning of honeybees by toxic chemicals and action to take when this occurs and the practical measures possible when prior notification is received.
5.12   An account of the expert services available to the beekeeper at national and county level.
5.13   The scientific names of the causative organisms of the aforementioned diseases and infestations.

6.0 Beekeeping.
6.1   Description of the various types of hive at present in use in the United Kingdom.
6.2   Description of the various frames in general use in the UK.
6.3   Definition and description of the concept of the "bee space".
6.4   The purpose of wax foundation within the moveable frame hive.
6.5   Two methods of wiring frames and embedding wire into foundation.
6.6   The various common methods of maintaining the spacing of frames in hives and the measurements of two recognised spacings.
6.7   An account of the queen excluder and the types in common use.
6.8   A detailed account of how to commence beekeeping, including the acquisition of bees, sources of equipment, costs and any precautions necessary when acquiring bees or equipment.
6.9   The criteria to be observed when moving colonies of bees from one place to another (including optimum distance, vibration, temperature, ventilation, water supply).
6.10   The factors to be considered when siting colonies in a small home apiary.
6.11   A detailed account of the year's work in an apiary including migratory beekeeping.

6.12   The principles of feeding a colony of honeybees.
6.13   The most common types of feeder in use.
6.14   The principles of supering.
6.15   The importance of supering as a factor in swarm prevention.
6.16   A detailed account of one method of swarm control and prevention.
6.17   Methods of taking and hiving a swarm.
6.18   Methods of making nuclei.
6.19   An account of the various uses to which nuclei can be put.
6.20   A detailed account of management of nuclei and swarms to turn them into productive colonies.
6.21   A detailed account of methods of uniting colonies and the precautions to be taken.
6.22   An account of the methods of dealing with laying workers.
6.23   A simple method of rearing a small number of queens.
6.24   The symptoms of queenlessness and how this may be confirmed.
6.25   Methods of queen introduction by cage, uniting or direct action, itemising necessary precautions.
6.26   The problem of robbing and methods used to avoid it or to terminate it once it has started.
6.27   The clearing of bees from supers, by the "shake and brush" method and by using clearer boards.
6.28   How to prepare and protect colonies for the winter period.
6.29   The damage to colonies caused by mice and other pests.
6.30   Methods of providing a suitable water supply for bees within the apiary.
6.31   An account of the management of colonies for the production of sections and cut comb.
6.32   An account of the preparation of sections and cut comb honey for sale.
6.33   Suitable methods of decapping super combs.
6.34   The principles of honey extractors, both tangential and radial.
6.35   Methods of straining small quantities of honey and its subsequent storage.
6.36   Methods of small scale bottling and preparation of honey for sale,
6.37   Details of legal requirements for labelling and sales of home produced honey.
6.38   Methods of storing comb with particular reference to prevention of wax moth damage and sterilisation against Nosema.
6.39   Wax moth damage to stored comb.
6.40   Small scale methods of recovering beeswax from both comb and cappings.
6.41   A method of preparing home made wax foundation.
6.42   The possible effect of stings and be able to recommend a suitable first aid treatment.

APPENDICES

A1 - Legislation applying to the supply and sale of honey in UK.
A2 - Wag-tail dances.
A3 - Average colony population cycle.
A4 - Anatomy and other diagrams.
A5 - Colony inspections (timing).
A6 - Recommended books.
A7 - Typical examination questions.
A8 - Manipulating a colony of honeybees.
A9 - Consumption of stores during winter.
A10 - Bailey frame change.

THE INTERMEDIATE EXAMINATION

- A candidate for the Intermediate Examination should thoroughly understand A1 to A3 and be able to draw all the anatomy diagrams shown in A4 quickly and from memory.
- The Intermediate Examination is held twice per year; usually the last Saturdays in March and November except when Easter clashes with this arrangement.
- Application should be made 6 weeks prior to the examination date through your County Examination Secretary.
- Any candidate for the examination must have previously passed the Basic Examination.
- The pass mark for the Intermediate Examination is 70% and the results are promulgated as grades as follows:

GRADE A+	> 85%
A	80 to 84%
B	70 to 79%
C	60 to 69%
D	50 to 59%
E	< 50%

- The Robert Hammond Prize [£25] is awarded once a year by the BBKA for the best result from the March and November examinations providing there is a result of a high enough standard.

SYMBOLS USED IN THE NOTES

>	greater than
<	less than
=	equal to
≈	approximately equal to
∝	proportional to
→	becomes
∴	therefore
k	kilo
μ	micro
n	nano
π	pi = 3·142
α,β,γ,θ	Greek letters denoting various angles
⊙	sun
☐	hive
▽	source of nectar
c.	circa (= approximately)
d	day
w	week
m	month
y	year
F	Fahrenheit
C	Centigrade
RH	Relative humidity
O_2	oxygen
CO_2	carbon dioxide
H_2O	water
Hz	Hertz (= cycles per second or cps)
cf	compared with

0.0 PREFACE

0.1   These notes were originally written when we were studying for the Intermediate Examination. Having been persuaded to make them available to other students studying on their own, we have edited and brought them up to date for publication. They were not initially intended to be a text book, only an aid for self study concentrating only on those topics in the BBKA Intermediate syllabus. Because of this, no index seemed necessary and references have been omitted; we make no apology for this, we may have to get down to it at some later stage.

0.2   There are a few problems for those studying for and attempting the Intermediate Examination. We believe now that there are many who have passed their Basic Examination and are 'put off' attempting the Intermediate because of the difficulty and cost of obtaining the books contained in the BBKA recommended book list. We consider that these notes, together with only three text books, should be sufficient for examination purposes. The notes contain no photographs in order to minimise the cost of publication. A further area of difficulty is the amount of time and effort required to seek out all the necessary information because it is not concentrated in one volume; we hope that this problem is rectified, to some extent, with these notes. The last problem is the very high standard required in the examination. The old prospectus indicated that the examination is a 'searching test' of the candidate's knowledge of beekeeping. The pass mark is 70% We consider it to be a quantum leap from the Basic to the Intermediate standard and this should be realised by prospective candidates.

0.3   The three books which should be purchased are:

- **Anatomy and dissection of the honeybee**   **Dade, H.A.**
- **The hive and the honeybee**   **Dadant.**
- **A guide to bees and honey**   **Hooper, Ted.**

Further books, as recommended in the BBKA list, should be borrowed and read. These can be obtained from your beekeeping association library, either county or branch or both and/or from public libraries. Public libraries are very helpful and usually allow students to have books (approximately 4) on long term loan without renewal for approximately 3 months. For the affluent student, the books recommended in appendix 6 would be a worthwhile buy and are available through specialist booksellers.

0.4   We have spoken to candidates who have failed the Intermediate Examination and in many cases we have found that their knowledge is not lacking. It is their inability to commit their knowledge to paper in the required time frame. In other words, acquire the examination technique. To do this, read a question then during the following week read as much as you can about the detail required to answer the question, writing notes on the salient points. Forget about the topic for a couple of days and then answer the question under examination conditions; a blank sheet of paper, pen and watch to time 30 minutes exactly. Then go over it all again with notes and books and find out what you missed out. Do two of these a week from past examination papers for 20 weeks and more than likely you'll be up to examination standard. It is all a matter of self discipline.

0.5   After these notes had been converted from the original ones, which of course were in manuscript, to the typed version for printing, they were read by one or two beekeepers in Devon who thought that they would be valuable also for Basic candidates. As a result,

Appendix 8 was prepared with these candidates in mind. Much of the subject matter is at a standard higher than that required for the Basic Examination but nevertheless the notes do cover the whole of the Basic syllabus and in this respect should prove useful. Prospective candidates should obtain a Basic syllabus and select the parts of direct interest to them

0.6   We believe that the standard of beekeeping, in the parts of United Kingdom that we know well, is of a low standard and could be improved if more beekeepers were to apply themselves to the BBKA examinations, certainly up to Intermediate standard. We hope the following notes will help those who have been experiencing difficulty and make study easier for those wanting to advance in the craft of beekeeping. They could provide a useful aide memoire for those not contemplating any of the examinations.

JD & BD Yates.
Newton Ferrers, Devon. 1990.

1.0 NATURAL HISTORY OF THE HONEYBEE

1.1 The life cycle of the female castes and the drone.

1.1.1 Definitions:

- Metamorphosis - change in form by magic or by natural development or change (usually rapid) between immature form and adult state.
- Caste (zoological definition) - form of social insect having a particular function.

It should be noted that the most recent syllabus for the intermediate examination only recognises two female castes and the drone. We believe the question of two or three castes is a matter of opinion, there being arguments in favour of both. As three have been recognised for over 100 years, the change seems unnecessary and we have not changed our original notes.

1.1.2 The three castes:

- worker   –   from a fertilised egg (female).....................32 chromosomes
- queen   –   from a fertilised egg (female).....................32 chromosomes
- drone   –   from an unfertilised egg (male).................16 chromosomes

1.1.3 Stages in the life cycle:

	WORKER	QUEEN	DRONE
OPEN CELL:			
Egg	3d	3d	3d
Larva (4 moults)	5d	5d	7d
SEALED CELL:			
Larva/pro-pupa (1 moult)	3d	2d	4d
Pupa (1 moult)	10d	6d	10d
From egg to emergence	21d	16d	24d
AFTER EMERGENCE:			
Summer bee	6w	c.3y	c.4m
Winter bee	c.6m	ditto	-

Note that the above times can vary by a few hours before emergence due to variations in temperature of the brood nest. [c.= approximately, d = day, w = weeks, m = months and y = years]

1.1.4 Description of the stages in the life cycles.

Worker (before emergence):

1st day of egg	- vertical, stuck to the bottom of the cell and parallel to the cell walls
2nd day	- at an angle of c. 45°
3rd day	- horizontal, egg laying on the bottom of cell. Egg hatches after 3 days.
4th-8th day	- larva grows, moulting every 24 hours, until it fills the whole cell diameter. The cell is sealed on 8th day after the larva's last meal.
8th-21st day	- the connection between the ventriculus and the hind gut opens and the Malpighian tubules open into hind gut; excreta enters hind gut and is voided into cell. Larva changes position and stretches out the full length of the cell (head outwards) and spins a cocoon. Metamorphosis occurs and the larva changes to a pupa after 5th moult 3 days after sealing. The pupa is still white but of adult form. It completes development, slowly changing colour and emerges from its cell by nibbling the capping on day 21. The 6th moult occurs just before emergence.

Queen and drone:

Similar but with the different timings shown above.

After eggs hatch, the larvae are immediately capable of eating food:

Workers and drones - are progressively fed with brood food for first 2-3 days, then a mixture of brood food, pollen and honey. No food is consumed during the pupal stage. Queens - are fed by mass provisioning with royal jelly throughout the the larval stage. Queens are generally over fed and excess can usually be seen in the cell after emergence.

1.2 Functions of the three castes in the life of the colony.

1.2.1 Worker:

1st to 3rd day	- Cell cleaning and brood incubation.
4th to 6th day	- Feeding older larvae (brood food + honey + pollen).
7th to 12th day	- Feeding young larvae (brood food only).
13th to 18th day	- Processing nectar into honey, wax making, water evaporation and pollen packing.
19th to 21st day	- Guarding and starting to forage.
3rd to 6th week	- Foraging for nectar, pollen, water and propolis.

Note that these times are approximate and older bees can revert to their earlier duties if required by the colony. Other duties include ventilation, humidity and temperature control, etc.

1.2.2 Drone:

Up to c. day 12	- Generally confined to the hive except on fine days for cleansing and orientation flights.
12th to 14th day	- Mature and ready to mate (his sole function).
Autumn	- Driven out of the hive to die.

1.2.3 Queen:

1st day	- Seeking rivals and killing them.
3rd to 5th day	- Orientation flights to locate the hive.
1st to 3rd week	- Multiple mating flights
Up to 3 to 5 years	- 3 or 4 days after mating the queen starts to lay. Thereafter, she is solely egg laying and producing pheromones for colony cohesion.

Note that up to the time the queen starts to lay, the times are not precise and are very much dependent on the weather. A queen that has not mated satisfactorily within c.20 days usually becomes a drone layer and is said to be stale.

1.3 Parthenogenesis.

1.3.1 Definition: reproduction without fertilisation (as from germ cells and in lower plant life). Word derived from two Greek words meaning virgin birth. Note that the queen is fertilised but has the ability to lay either fertilised or unfertilised eggs.

1.3.2 The theory of parthenogenesis in the honeybee was propounded by Dzierzon in 1845 i.e. males or drones are produced from unfertilised eggs. Dzierzon's theory has required a minimum amount of modification during the last 100 years.

1.3.3 The queen has the ability, at will, to fertilise an egg before laying. Koeniger in 1970 showed that the queen uses her front legs to measure the cell in order to determine whether the egg should be fertilised or not.

1.3.4 In very rare cases unfertilised eggs can give rise to females. Onions found this to be the case in South African bees in 1912 (Apis mellifera capensis).

1.4 Caste production in female honeybees.

1.4.1 Caste determination: both the queen and the worker are derived from a fertilised egg. Therefore differentiation between queen and worker cannot be due to any genetic differences and must be due something else (e.g. feeding).

1.4.2 The queen is fed on royal jelly (a glandular secretion from the hypopharyngeal and mandibular glands of the worker bee) throughout the whole period from hatching of the egg to the propupa stage of development. A plentiful supply of royal jelly is available at all times for the larva. The larva continues to feed on the same diet after the cell is sealed.

1.4.3 The worker is fed on brood food for the first 2-3 days after the egg hatches, then honey and pollen are added to the brood food up to the sealing of the cell. Note that no food is left in the cell as in the case of the queen.

1.4.4 Analysis of larval food has given variable results:

- In 1888 Planta postulated that differing foods determined the female caste (queen or worker),
- In 1943 Haydak postulated that it was not the type of food that determined caste but the amount consumed. Experiments to test this theory have not been fully conclusive.
- In 1956 Weaver carried out feeding experiments to confirm Von Rheim's work in 1933 that there was a 'fugitive' substance in the larval food.
- In 1961 Jay disputed the interpretation of Weaver's results.

It is clear that, at present, the exact mechanism determining caste of the female is not fully understood and more work is required to answer the problem fully.

1.4.5 The only other differences are in the cell (ie. worker and queen cells); the size and orientation have been shown to have no effect on caste determination.

1.5 Detection of a drone laying queen & the causes for this failure.

1.5.1 The visual signs:

- Unmistakable worker cells with drone cappings (raised).
- Presence of a queen (actually seen).
- Drones produced are small and abnormal (stunted).

1.5.2 During the season:

- Queen produces small areas of drone brood in the middle of large patches of worker brood.
- As the season progresses, worker brood becomes less and drone brood increases.
- Because some worker brood remains, it is clear that a queen must be laying.
- Eventually there will be nothing but drone cappings. At this stage the colony will be reasonably large.
- Drones are smaller and the abdomen stunted.

1.5.3 In the spring:

- Very difficult to detect at the first examination of the colony ie. a small colony with one or two frames of drone brood only (no worker).
- Is it a drone laying queen or laying workers?
- If a queen (drone layer), the laying pattern will be orderly i.e. compact patches of brood with very few empty cells.

1.5.4 Possible causes for a queen becoming a drone layer:

- Shortage of sperm - inadequate mating or due to age.
- Physical inability of queen to fertilise eggs correctly.
- Genetic fault.

1.5.5. Treatment:

- requeen or
- unite after removing old drone laying queen.

1.6 Detection of laying workers and description of why they occur.

1.6.1 Detection of laying workers:

- Drones in worker cells (typical raised domes).
- Drones produced in this way are small and abnormal (stunted).
- Laying pattern is scattered and haphazard (cf. drone laying queen which is compact and orderly.
- Colony endeavours to build charged queen cells (note: this can happen with drone laying queen but is unusual).
- Workers generally lay more than one egg/cell.

1.6.2 In the absence of the queen and brood there is an absence of pheromones from the queen and from the brood. These pheromones, particularly that produced by the queen, inhibit development of a worker's ovaries. Workers' ovaries develop and some workers start laying after about 21 days in the queenless state.

1.6.3 Cause for colony having laying workers:

- Queenlessness.
- Inability to produce emergency queen cells (no fertilised eggs).

1.6.4 Treatment: it is generally agreed that little can be done except to shake the colony out near a strong stock and let them take 'pot luck'.

The following points are pertinent:

- Difficult (impossible?) to requeen; a colony usually kills an introduced queen.
- Bees are mostly old and not much use to another colony.
- If they are united to a queenright colony it has been found that there is the likelihood of them killing the queen of the colony to which they are united.
- Experiments conducted in France in 1989 on the introduction of queens to colonies of laying workers by dipping the queen in royal jelly and water (70% and 30% respectively) are claimed to be a successful treatment.

1.7 Sexual reproduction of the honeybee; aerial and multiple mating.

1.7.1
- Queen mates on the wing between 5-20 days after emergence.
- If she has not mated in 3-4 weeks she is no longer capable of mating properly and is known to be 'stale'. Sperm cannot migrate through the duct leading to the spermatheca.
- Bees in a colony with a virgin queen become more and more aggressive to her until she mates.
- This aggressive behaviour may be responsible for driving out the virgin queen before she is too old to mate and become stale.
- After mating, bees are very attentive to the queen, grooming her and forming a court around her.

1.7.2 Drones often have collecting points where they tend to congregate. Such a congregation point attracts drones from a wide area ensuring drones of varying strains and thus minimising inbreeding. It has been observed that mating only occurs at heights of greater than 30 ft. and less than 90 ft. above the ground. Also, the height of mating is inversely proportional to windspeed.

1.7.3 On the mating flight:

- Queen flies to the level of the drones.
- Drones are attracted by one of the pheromones from the mandibular and other glands of the queen and form a 'comet tail' flying behind the queen. The drone detects the queen at about 50 metres by pheromone and sees the queen at about 1 metre.
- First drone, stimulated by another pheromone, mates with the queen and drops to the ground to die.
- Further matings (5-15) occur on 2 or 3 separate mating flights.
- Mating continues until the spermatheca is full of sperm. After mating the queen has sufficient sperm to last her life of 3-5 years.
- The vaginal opening of queens returning to the hive after mating flights often contain the male genitalia which is removed by the bees inside the hive.

1.7.4 Mating flights normally occur in good weather when there are plenty of drones flying (say noon to 4 pm) at temperatures 20°C and greater. High winds discourage mating flights. Average length of time of mating flights have been observed to be c.20 mins. in April which decreases to c. 12 mins. in June.

1.7.5 The queen starts to lay c. 2-4 days after mating is complete. Egg laying is often erratic when the queen starts to lay and more than one egg per cell occurs. This phenomenon generally soon disappears.

1.8 Description of the work undertaken by worker bees.

1.8.1 The work undertaken by the worker bee is generally dependent on the age of the bee and the development of various glands:

- 0 - 3d.   - House (hive) cleaning, eg. cells for q. to lay in.
- 3 - 9d.   - Feeding larvae / nursing.   *
- 9 - 18d.   - Ripening honey.   *
                 - Wax making / comb building.   *
                 - Ventilation / evaporation.
                 - Temperature control.
- 18 - 21d.   - Guarding / defence.   *
- 3 - 6w.   - Foraging.

* - these activities require a glandular activity in the bee; 1st the hypopharyngeal and mandibular glands develop for producing brood food / royal jelly and enzymes for processing nectar into honey, 2nd the wax glands become operative at about 12d. and finally the sting produces venom at about 18d.

It should be noted that the bee works for 8 hours, patrols for 8 hours and rests for 8 hours, although these activities are not performed in 8 hour stretches.

1.8.2 Note rule of three!

> 3 castes q., w., d.
> Egg 3d to hatch.
> Egg to worker emerging 3w.
> Duty as house bee 3w.
> Field bee 3w.
> Life of drone c. 3m.
> Life of queen c. 3y.

1.8.3 House cleaning. This includes:

- Cell cleaning - removal of excreta, larval moults and then polishing cells ready for laying.
- Hive cleaning - removal of dead bees and debris from the floor. These are menial tasks undertaken by the youngest bees with no experience of other duties; they are performed instinctively and start more or less immediately the bee emerges from its cell.

1.8.4. Feeding / nursing.

• Feeding older larvae	3 - 6d.
• Feeding young larvae	6 - 9d.
• Capping brood cells	9 - 12d.

At 3d. old the hypopharyngeal glands start to become active and the nurse bees take up feeding duties for about 6d. (3-9d. old). At about 9d. old the wax making glands become active. Note very young and very old bees do not secrete wax. From about 9-12d old the nurse bees (now secreting wax) start brood capping duties. Note the colour of the cell cappings; they contain pollen mixed with beeswax to make them porous allowing the larvae/pupae to breathe. Cells filled with honey are capped with pure wax.

1.8.5 Processing nectar into honey.

There are two distinct processes, namely:

- Chemical change in the the sugar content of the nectar.
- Physical change whereby the surplus water is evaporated in order to ripen it, after which it is sealed in cells with pure beeswax. Note honey is hygroscopic and the capping excludes air and hence prevents the absorption of water from the atmosphere.

Chemical change: nectar contains varying amount of sucrose* which is converted into glucose and fructose by the enzyme invertase derived from the hypopharyngeal glands. Note that some books state that invertase is derived from the postcerebral and thoracic glands and the chemical conversion starts on the foragers' flight back to the hive. The load is passed from a forager to a house bee for ripening. The house bee swallows the nectar and regurgitates it every 5-10 secs. for about 20 mins. and then the nectar is hung in droplets in empty cells or placed in partially filled cells to dry.

Physical change: nectar contains approx. 20-40% sugar (typically 30%) when first brought into the hive. The ripening process (regurgitation) brings it up to c. 45% sugar concentration when it is put into the cells. After evaporation it is typically 80% sugar content. the water content is typically 17% and the other 3% constitutes acids, mineral salts, etc.

* Sucrose is a disaccharide, fructose and glucose are mono-saccharides; honey also contains small quantities of higher order sugars.

1.8.6 Wax making/comb building.

- Wax secretion generally occurs in bees 12-18 days old at relatively high temperatures 33-36°C (91·4 - 96·8°F). The wax is secreted in small flakes from 4 pairs of wax glands on the last 4 visible segments on the ventral side of the abdomen (colloquially known as the waistcoat pockets), ie. on the sternites A4 - A7.
- Large honey consumption is required in order that the bee may produce wax. In the literature various estimates are given; however 8lbs of honey for 1lb of wax seems to be a realistic mean.
- When building comb, workers gorge themselves with honey and hang in festoons for c. 24hrs. before the wax secretion and building process starts.
- A wax scale is removed by one hind leg and transferred to the mandibles by the two fore legs. The wax scale is thoroughly masticated before fixing to the comb and moulding it in place. When it is first deposited it is spongy and flaky and is later manipulated again making it smoother and more compact. Removing, masticating and fixing one scale takes about 4 minutes. It is not clear whether any secretions from any of the glands are used in comb building (mandibular, salivary, etc.). Some books indicate that the mandibular glands are used.
- Bees can detect gravity (sensilla at the petiole between the head and the thorax) and the festooning chains (catenaries) play an important role in the parallelism of the combs.
- Queenlessness and bright light inhibit the bees to build comb and secrete wax.
- According to Dadant, the thickness of the wall of newly built comb is approx. 0·0025" thick and in naturally drawn comb without the use of foundation, the base is 0·0035" thick; Hooper gives 0·006" for the cell wall thickness.

1.8.7 Evaporation.

- The sugar content of nectar varies considerably depending on temperature, humidity. sunshine, wind speed and direction, etc. If the incoming nectar contains 40% sugar then there is c.60% water. After manipulation by the house bees, the ripened nectar will contain c.15% less water, ie. 45% water. Ripe honey contains c.20% water; the difference between 45% and 20% is due to evaporation as warm air is passed over the combs by fanning bees.
- Nectar is first manipulated by the bees (gorging and regurgitating), causing both a physical as well as a chemical change before it is hung up in droplets in empty or partially filled cells (ie. largest surface area). It is spread over a large area of comb and later gathered up and concentrated into a smaller area of comb. For this reason it is very important to provide adequate comb space during a nectar flow.
- On average it takes about 4/5 days to ripen nectar from 60% water content to honey of c.20% water content.
- During the journey back to the hive, the water content remains virtually unchanged.
- Evaporation is dependent on:
    - storage cells (space) available
    - temperature (directly)
    - humidity (inversely)
    - ventilation
- There is a need for a continuous stream of air from outside to inside as the air inside becomes saturated. The RH inside the hive varies from 20%-80%. In the brood nest it is

fairly constant, 35%-45%RH.

- As an example during a heavy flow with a strong colony c. 2·5Kg (5.5 lbs) of water are evaporated in 24hrs. or 50% of the gross gain per day. c.⅔ of this total loss occurring during the day.
- For each field bee (forager) there are approx. two bees in the hive for house duties which include the ripening and storage of honey. The importance of good hive ventilation is clear; it being important to allow top ventilation through the crown board and roof ventilators as well as the hive entrance at the bottom.

1.8.8 Ventilation.

- Ventilation of the hive is always necessary in order to expel CO_2, to expel water vapour when the colony is ripening honey and for cooling purposes in hot weather. This can be achieved by the bees but assistance can be given by the beekeeper so that the colony is put under minimum stress.
- When the temperature of the brood nest (93-95°F) tends to rise above its normal limits, there is a colony response. The bees sense the need for ventilation to control both the temperature and the humidity (35-45%RH). Temperature control is necessary to incubate the brood and control of RH is required so that brood food and royal jelly, when present, does not dry out.
- By evaporating water from nectar or from water actually brought into the hive, the heat is 'used up' thereby cooling the hive. When water evaporates there is a drop in temperature (known as the latent heat of evaporisation). The bees do this by fanning; it is a very economical and efficient air conditioning system.
- Fanning: bees collect to one side of the entrance and face inwards and fan vigorously creating an outward flow of air. Others are doing the same thing further in on the floorboard. There can be up to hundreds fanning and the draught created can easily be felt with the hand near the entrance. If the conditions become extreme, a further group will start fanning on the other side of the entrance but facing outwards and setting up in inward current of fresh cool air. Fanning is normally undertaken by older bees who are muscularly strong to perform this task.
- The beekeeper can assist in hot weather by:

  - providing sufficient space within the hive,
  - providing shade over the hive at noon and/or pm,
  - ensuring the crown board feed hole is open and the roof ventilators are not blocked,
  - keeping undergrowth trimmed around the hives,
  - in extreme conditions:- staggering supers, off setting crown boards and roofs, raising hive above the floor (say 1").

Note that late in the season to prevent robbing the hive must be beetight and only have one entrance which can be guarded.

- When bees are being moved the temperature increases very rapidly due the disturbance and travelling screens are essential, particularly in summer.
- Ventilation in winter: The degree of ventilation required for successful wintering is not agreed among the experts and the literature can be confusing on the subject.The amount is determined by the sizes of the openings top and bottom and the convection currents (warm air rising around the cluster). the variables are:

  - size of entrance (nb. mouse guards),

• size of top ventilation	- raised crown board (matchsticks)
	- Morris board
	- crown board completely removed
	- size of roof ventilators.

It is generally agreed that greater ventilation is required in warmer and damper regions cf. colder and drier regions. However, it should be noted that about 4 gallons of water have to be dispensed with by evaporation due the consumption of c. 30lbs of winter stores.

1.8.9 Temperature control.

• Individual bees have no means of controlling their body temperature and quickly assume the ambient air temperature (they are said to be poikilothermic). Their activity, both physical and physiological, quickens with a rise of temperature and slows as it falls. This automatic effect of temperature on metabolism may serve to stimulate equally automatic social actions to adjust the temperature of the hive. It should be noted that the bee is capable of sensing temperature, if it could not do so it would be unable to survive.

• Temperature is lowered by:

    • fanning at entrance and inside the hive,
    • water evaporation,
    • dispersion through the hive (as opposed to clustering),
    • clustering outside the hive entrance.

• Temperature is increased by:

    • muscular activity (thorax muscles),
    • clustering tightly,
    • manipulating a colony or moving a colony (muscular activity).

• Activity temperatures:

    • all activities occur between 10°-38°C (50°-100°F),
    • brood nest ≈ 35°C (95°F),
    • unable to fly at 10°C (50°F) *,
    • becomes immobile at 7°C (45°F),
    • clustering starts at 14°C (57°F),
    • thorax T 20°-36°C (68°-97°F) normally 29°C (84°F).

* Very often bees can be seen flying at air temperatures lower than this; the actual temperature of the bee is therefore above 10°C. Water collecting in the spring can be a very hazardous occupation for the honeybee; it must not allow its body temperature to fall below the critical 10°C while it is taking water or it can never return to the hive.

• The lower the temperature, the tighter the cluster (physically smaller) thereby providing a smaller surface area and less heat loss. At very low temperatures the bees bury their thoraces in the cluster and spread their wings (also to reduce heat loss) with abdomens out. the connective cluster under these conditions merges with the main cluster.
• The old adage that bees never freeze to death, only starve to death, is very accurate. With an

adequate supply of stores they can survive very low temperatures by generating sufficient heat in the centre of the cluster to maintain the outer surface of the cluster at just above 7°C. This stops the bees on the outside from becoming immobile and falling off. Note that, contrary to popular belief, the bees in the cluster do not continually change position to keep warm.

1.8.10 Defence.

• Defence generally occurs at the hive or within a few metres of it. The defensive vigour of a colony depends on the genetic 'make up' of the strain, some bees being much more aggressive than others.
• No guard bees are likely to be found at the entrance of a colony during a nectar flow. Conversely, during times of dearth, many guard bees will be seen.
• Guard bees exhibit a typical stance - standing on their 4 rear legs, forelegs raised and antennae outstretched. If some become alarmed, they open their mandibles and spread their wings ready for attack.
• Each guard bee 'patrols' a particular area. They check incoming foragers and challenge drifting bees by touching with their antennae (1-3secs.).
• Robbers (both bees and wasps) have a distinctive flight noticeable to guard bees who will always attack in defence rather than challenge first.
• Stinging is a defensive mechanism, not an attacking one. In an undisturbed colony less than ½% of the bees in a colony are likely to sting (200 in 40,000).
• Guard bees are sensitive to:

  • vibrations,
  • visual stimulus (fast movements),
  • odours (animals, humans and pheromones).

Once the guard bees have been alerted, they will fly round the area outside the hive, the distance they guard depending on the strain of bee. The Africanised bee in S. America will guard at distances greater than ½ mile. Many bees in UK (usually bad tempered stocks) will follow for a few hundred yards. A few yards is typical for a reasonably tempered colony.
• Bees from the same colony have the same smell due to them having the same diet resulting from the food sharing and transmission among all the bees in the colony. For these reasons Bro. Adam believes there is a hive odour. Dr. Colin Butler on the other hand maintained there is a colony odour which is genetically produced. Whatever the reason it is generally agreed that guard bees can recognise bees from another hive whether they be drifters or robbers.
• Colonies working the same flower (eg rape) would be expected to develop the same hive odour and recognition would be through colony odour. However under these conditions, where a flow exists, there are usually no guard bees present and it is clear more work is required on colony versus hive odour theories.
• Stimuli that elicit stinging behaviour are:

  • exhaled breath,
  • smell of hair, leather and many cosmetics,
  • violent vibrations and bumps,
  • rapid movements,
  • most important - alarm pheromones (bee venom and isopentylacetate from the sting and 2-heptanone from the mandibular glands).

• Guard bees are usually 18d. old prior to becoming foragers. When the guards are alerted and pheromones are distributed around the hive, foraging tends to stop and the foragers become guard bees; many more bees are noticeable at the hive entrance under these conditions.

1.8.11 Foraging.

• In a well balanced colony in the season c.⅓ worker bees are foragers, the other ⅔ are house bees. Workers normally start foraging at c. 21 days old and continue until they die (away from the hive) c. 3 weeks later. They forage only in the daylight hours and in favourable weather; T≈55°F (13°C) and above, flying 6-10ft. above ground in winds below 15mph (24km/h). Flying at 15mph, they consume honey at a rate of 10mg/hr.

They forage up to c. 2·5 miles from the hive. They forage for nectar, pollen, propolis and water. When T=43°C (109°F) and above they only forage for water. A small number of the foragers (about 2%) act as scout bees, a very important activity; however it is virtually impossible to determine the exact number.

• Foraging is stimulated by:

   • presence of the queen and brood (both produce stimulating pheromones),
   • the needs of the colony indicated by food exchange in the colony and the speed foragers are unloaded by house bees,
   • scout bees dancing on the combs providing information on distance, direction and quality of the source. The most stimulating dances attract the most foraging activity.

• Foraging bees collect nectar, pollen, water and propolis.
   • **Nectar**: this is collected from the nectaries and extra floral nectaries of suitable flowers providing they are yielding nectar and the bee's proboscis is long enough to reach it.
            - Nectar is sucked up by the proboscis, the average load being 40mg. and the maximum load c. 80mg.
            - The returning forager unloads by passing the nectar load to a house bee for manipulation and subsequent ripening.
            - The speed of unloading by the house bee indicates the colony's needs (note that high sugar content is preferable to low sugar content).
            - Foragers are constant to a particular species of flower and will rest in the hive if it is not yielding at a particular time.
            - The number of trips per day is an average of 10 and generally range from c. 7-13 trips.
            - The average time per trip is c.1 hr., half this time is spent flying and the other half collecting and unloading. When unfavourable conditions prevail, trips as long as 3-4hrs. have been reported.
            - Number of flowers visited per trip to obtain a full load is very variable (50-1000) and depends on all the variables associated with the secretion of nectar.
            - It should be noted that bees also forage for honeydew, the sugary secretion of many species of aphid.
   • **Pollen**: this is collected from the anthers of flowers either intentionally or by chance when foraging for nectar. It is collected by the plumose hairs covering the exoskeleton and transferred to the corbiculae by the 3 pairs of legs.
            - The average load (weight of both pollen pellets) ranges from 11-29mg.
            - The pollen collecting trips are completed more quickly than nectar collecting trips. Range is reported to be 3-18 minutes.
            - The pollen forager returns to the hive and deposits the load directly into a cell adjacent to the brood nest without assistance from house bees. Later, house bees come along and pack the pollen loads into the cells and finally, for storing, they are sealed with a layer of honey and wax.
            - Open brood provides a pheromone which induces foraging for pollen in addition to

queen substance.
- The number of pollen foragers is controlled by the needs of the colony and the number of cells prepared by the house bees for pollen. It has been reported that they can vary from a few % to as high as 95% of the total foraging force.
- The number of foragers collecting mixed loads of both nectar and pollen is c. 3% of the foraging force.

• **Propolis**: this is a resinous exudate from the bark or buds of trees which is collected by the bees for filling cracks in the hive, reducing openings (eg. entrances), smoothing the interior of the hive, varnishing the interior of brood cells, strengthening comb attachments and embalming objectionable objects too large to remove from the hive. It is collected with much difficulty and transported back to the hive on the corbiculae.
- It takes a long time to collect and has to be unloaded with assistance from house bees, again taking a long time (1-2hrs.)
- Some strains of bee collect more propolis than others (eg.Caucasians) and therefore have a larger % of the foraging force collecting at the expense of nectar and pollen.
- In times of dearth, foragers will seek and use alternatives (eg. tar from roads).

• **Water**: this is the only item which is collected and not stored in the hive. It is collected in the same way as nectar with the proboscis and transported to the hive in the crop.
- Water is required for diluting honey so that it can be metabolised, diluting honey for brood food for larval feeding, for humidifying the brood nest and for cooling the hive in very hot weather.
- The average load is 25-50mg. and the foraging trip is very short, most are completed in under 10 minutes.
- The average colony in the spring requires c.150g. per day and a strong colony in hot drought conditions requires c. 1kg.per day.
- Reception by the house bees indicates the colony needs; if unloading takes longer than 3 mins. then water collection activity ceases.
- Water foragers mark favourable sites with Nasonov gland and fanning.
- At times water is stored, not directly in the hive but in the crops of the receiving house bees (reservoir bees); this happens when supplies are not readily available.

1.9 How the bee orientates to the hive.

The dictionary defines 'orientation' as applied to birds and insects as a faculty of finding their way home from a distant point or place.

• Orientation in the case of the honeybee involves:

- • the use of landmarks (in close proximity to the hive),
- • position of the sun (both altitude and azimuth),
- • ability of the bee to detect polarised and ultra violet light.

Note also the ability of the bee to fly round obstacles to and from a food source.

• Young bees (during the time they are house bees) take short flights around the hive noting the landmarks in the very close vicinity (eg. large stones, a bush, a tree, etc.).
  • Occasionally much activity may be noticed at the hive entrance; these are young bees on initial orientation flights (popularly known as play flights). It is curious that quite a large number do this simultaneously. Successive flights are longer and further afield and occur about the time the bee is converting from a house bee to become a forager.

- By the time the bee starts to forage it has a very precise knowledge of its immediate environment (say half mile radius from the hive). Moving the hive only 2 or 3ft. disorientates the homecoming bees for a few hours until they re-adjust. It is for this reason that hives may only be moved 3ft. or 3mls. (the normal limit of their foraging range). Experiments conducted by the authors seem to indicate that when a hive is moved, the bee can only remember its old surroundings for c. 2 weeks; after this time it has completely forgotten its old hive vicinity.
- With no immediate local landmarks, young bees tend to drift into other hives. Young bees drift more than older bees. Drifting is more pronounced when the bees have been confined in the hive for long periods (eg. winter); nb. the 2 week memory. It is therefore important when siting colonies in an apiary to have:
  - hives painted different colours,
  - hives facing in different directions,
  - plenty of local distinctive landmarks.

  Note that drifting can spread disease between colonies.
- All the above are visual; there is one other local orientation system frequently used (particularly after disorientation) and that is scent fanning, using the Nasonov pheromone, at the hive entrance.

• The honeybee only flies during daylight hours using its eyes for navigation; it is not equipped for night flying. The ocelli (simple eyes) are incapable of focusing an image and are used solely for measuring light intensity. The compound eyes do produce an image (albeit a poor one by our own standards); they are however capable of detecting polarised light and ultra violet light (UV). It is this capability which not only enables the bee to return to its hive but allows it to communicate the direction and distance to a source of forage to other bees in the hive.

- The sun is an extremely powerful source of UV being a maximum in the sun's direction. At other angles to the sun, the strength of the UV decreases as the angle increases. If the sky is overcast and completely covered in cloud, the UV radiation still penetrates the cloud layer and the bee can detect it. The strength of the radiation is still a maximum in the direction of the sun; therefore the bee always knows in what part of the sky the sun is to be found.
- The light from the sun becomes polarised as it passes through space and the earth's atmosphere. The plane of polarisation being different in any particular direction (altitude above the horizon and azimuth). The compound eye of the bee can detect the plane of polarisation. Knowing the plane of polarisation and the direction of the sun, it automatically knows the angle between the two and can communicate this angle to the other inmates of the hive. The plane of polarisation defines the forage source which the bee flies to and it returns on a reciprocal course by the same means.
- Bees have another inbuilt mechanism; they can allow for the movement of the sun in the sky (15°/hr.) when calculating the angle to fly or to dance on the combs.
- When they reach the source of forage the foraging bee goes from flower to flower and finally returns to its starting point at the forage before setting a final course back to the hive. How the zig-zag course in the forage is calculated is still a mystery, although it has been proved that the sun is required.

1.10 Communication dances of worker honeybees.

• The main dances for communicating nectar and pollen sources, discovered by von Frisch, are:

> • Round dance   -   sources up to 100m
> • Wagtail dance   -   sources over 100m

• Round dance: contains little or no information except that the source is close to hand (within 100m). Bees (newly recruited foragers) responding to this dance search in all directions from the hive. This dance is most apparent if wet supers are replaced on a colony after extraction during daylight. The colony very quickly (a few minutes) goes into a state of agitation with many bees 'milling around' the neighbourhood of the hive looking for the source which of course is on the hive above the brood chamber. This important 'deficiency' in the bees' communication system is the reason why wet supers should ONLY be returned to the hive after dark when the bees cannot fly.
• Wagtail dance [see appendix 2]: provides very precise information on the direction and distance from the hive to the source of forage and the time it is available.
> • Direction is given as an angle between the food source and the sun. This angle is translated as the same angle between the vertical and the 'wagging direction' on the comb. The bee can determine the vertical by gravity sensors (between the head and thorax). The top of the comb always represents the sun. If the source, when viewed from the hive, is to the right of the sun, then the dance on the comb will be to the right of the vertical. Similarly for sources to the left of the sun, the dance will be to the left of the vertical.
> • Distance information is given by the number of 'straight runs' (centre of the pattern) every 15 seconds as follows:
>> 100m   = 9-10 runs/15secs.
>> 600m   = 7   ditto
>> 1000m   = 4   ditto
>> 6000m   = 2   ditto
>
> Wenner (1962) discovered that during the wagtail portion of the dance, a series of sound blips are made ($\approx$250Hz) which are inaudible to the human ear. The number of blips also correlate with distance. It is generally agreed that the number of runs per 15secs. is probably the most reliable indicator of distance.
> • Time the forage is available is the time the bee is dancing on the comb. It should be noted that bees have the ability to 'remember' time (when the sun is in a particular direction).

• Other types of dances (which are poorly understood):

> • Alarm dances (eg. poisoned food) - spirals or zig-zag.
> • Cleaning dances - stamping legs + swinging body side to side.
> • DVAV* ('joy') dances - front legs on another bee + 5/6 shaking movements.
> • Massage dances - starts by bending head in curious way (sickness)
> • Vibration dances - just before a swarm departs.

> * = Dorsal - ventral - abdominal - vibration

1.11 Annual population cycle of the honeybee colony.

• The honeybee (Apis mellifera) is now acknowledged to have its origins in East Africa. Development took place in two directions, one to the east producing the eastern bees and the other to the north producing the western bee, apis mellifera. Other races in Africa also evolved (eg. apis mellifera capensis - the Cape bee of S. Africa). Over the evolution period of c. 20 million years the flowers (angiosperms) also evolved, the honeybee and the flowers being complementary to one another. Both the honeybee and the flowers which they depend on for their food, have adapted themselves to the climate they live in.

> • Tropics: florescence has a seasonal variation (2 peaks of plenty and 2 peaks of dearth).
> • Sub-tropics: no seasonal variation (almost continuous florescence).
> • Temperate zones: extreme seasonal variation (1 peak of plenty and 1 peak of dearth):
> > - In N.temperate countries there are no flowers (or very few) from October to March and the colony survives on stores collected between April and September.
> > - The periods of plenty during April to September are generally very short (nectar flows only lasting a few weeks).
> • It is important to note that the population of the honeybee colony has adapted, during its evolution, to survive under these conditions.

• Reference should be made to the graph showing the variation in colony population over a complete year [appendix 3]. It is important to understand the variations for the successful management of a colony. Points of interest on the graph are as follows:

> • Brood = adult bees twice per year.
> • Brood > adult bees from Feb. to April. This is a very critical time in the annual colony cycle [nb. the danger of chilling brood and not having enough adult bees to incubate this brood].
> • Brood peaks in early/mid June.
> • Adult bees peak end June/start July (3 weeks after brood peak); this is the time the main flow usually starts when the maximum foraging force is required.
> • After the main flow the population starts to decrease, rapidly at first (old foragers dying off) then more slowly as the winter bees (6 month life) start to appear in the colony.
> • The minimum adult population occurs c. end February (c. 5000).
> • The maximum population will vary from 40,000 to 60,000 depending on the fecundity and strain of the queen.
> • The population builds up on the 'spring flow' often using all the income and storing very little.
> • The maximum adult population stores very large amounts in a short time for winter (much less brood to care for).
> • The reduced population allows adequate reserves for winter.
> • Brood rearing ceases in late autumn and starts again after the winter solstice when the days start to lengthen.
> • There is a continual decrease in population throughout the winter so dying bees are not abnormal at this time. The healthy colony removes any that die in the hive.

It should be noted that the graph is a representation of average conditions and the local flora and climatic conditions will modify it accordingly. Similarly these local variations mean a peaky graph and not a smooth curve in practice.

1.12 Influence of local flora and weather on colony population.

• The general tendency of a colony is to build up its population on the nectar flow from the spring flowers and for the large population to gather large amounts for winter use from high yielding flowers in the summer. UK sources of bee forage in approximate order of flowering is shown below:

1 willow	12 turnip	23 charlock	34 sanfoin (late)
gorse	rape (winter)	fld. bean (spr)	willow herb
blackthorn	sycamore	rape(spring)	bell heather
plum	maple	white clover	red clover
cherry	h. chestnut	mustard	sea lavender
gooseberry	hawthorn	blackberry	dwarf furze
currants	holly	sw. chestnut	ling
dandelion	strawberry	runner beans	ivy
apple	fld. bean (wtr)	lime	wild clematis
kale	raspberry	wild thyme	
swede	sainfoin (early)	lucerne	

• The spring blossoms (approx. 1 to 13 above) provide pollen and nectar which is not usually as high in sugar content as the summer crops. Bees will always work the flowers yielding the greatest sugar content. It will be obvious that the bees have a greater amount of work to do on nectar with low sugar percentage cf high sugar percentage.
• The summer blossoms (approx. 14 to 39) tend to yield nectars with higher sugar content, due to the warmer sunnier weather required for nectar secretion. Sugar content of nectar can vary from about 5% to 70% depending on the species and weather.
• The late summer/early autumn flow (39 to 42): heather is a specialised crop needing specialised management and ivy is a bonus at the end of the season. The authors have established that normal sized colonies collect about 10 to 15lbs. of ivy honey most years in the south of England (some years more).
• When nectar starts to flow into a colony, the queen is fed more thereby inducing her to lay more; brood rearing increases. The increased brood produces pheromones that induce pollen collection by the foragers and plenty of pollen incoming also induces the production of royal jelly and brood food; a chain reaction in brood production within the colony. Conversely, a dearth of nectar tends to inhibit brood rearing and pollen collection.
• Nectar flow is dependent on temperature (air and soil), humidity, wind speed and soil type; a complex set of variables. Temperature is very important; the enzymes in the flower which make the nectaries secrete need to be above a minimum temperature (eg. in clover this is approaching 70°F for it to yield well).
• Inclement weather which confines the bees to the hive halts the collection of nectar and retards brood production. Again the converse encourages brood production and colony increase if the bees can fly well.

1.13 Description of queen substance and its influence on queen cell production.

• Queen substance is a pheromone* produced mainly in the mandibular glands of the queen. The queen's mandibular glands are large cf. a worker's which are smaller and those of a drone which are rudimentary. Queen substance was discovered by Dr.C.G.Butler in the 50s at Rothamstead. It is extremely attractive to worker bees and there are indications that it is also produced (in much smaller quantities) by glands on the queen's abdomen. The Koschevnikov glands in the sting chamber may also be another site of production. Although the worker bee is a female caste, its' mandibular glands do not produce queen substance; they do however produce an alarm pheromone plus other substances which may be used for working wax and a fatty acid for preserving brood food.

> * Definition - a pheromone is a chemical, secreted from an exocrine gland of an animal, that elicits a behavioural or physiological response by another animal of the same species and so acts as a chemical message.

• Queen substance from the mandibular glands contain (see section 2.4.4):

> • 9 oxodecenoic acid (9 ODA),
> • 9 hydroxydecenoic acid (9 HDA),
> • other substances (some which have been identified).

• The substance becomes spread all over the queen's body probably when she grooms herself 2 or 3 times per hour. This is transferred to workers feeding the queen who in turn transfer it to other workers and so on throughout the whole colony by food transfer and also by antennal palpatation of her abdomen, most noticeable in the court surrounding the queen.
• It has a profound effect on the worker bees in the colony:

> • it inhibits the development of worker ovaries,
> • it inhibits the building of queen cells in the colony.

• In order that this inhibiting effect takes place, it is necessary that each worker in the colony receives a minimum regular threshold level. Queens produce c. 5000µgm/day during their 1st year. During their 2nd year they produce c. 2500µgm/day and so on, halving each year (exponential law of decay). It will now be clear why young queens are less prone to swarm cf. old ones; with plenty of queen substance queen cell production is inhibited.
• Overcrowded colonies tend to prevent adequate food transfer throughout the colony and are then likely to build queen cells and swarm (ie. chemical communications break down).
• Similarly, a colony that has been queenless for about 5-6 hours will be effected and will start building queen cells (emergency cells).
• Lack of queen substance can be caused by:

> • an ageing queen,
> • a damaged queen,
> • overcrowding or congestion (inadequate comb space),
> • a physically small queen (scrub queen).

Note that reduction in queen substance is not correlated with egg laying.

1.14 Description of food sharing in a colony.

• Food exchange [trophallaxis] is one of the most frequently observed behaviour patterns in a honeybee colony. It is going on 24 hours per day, day in and day out. Nectar or honey is passed from one bee to one or more receiving bees and is the prime mechanism for the exchange of pheromones within the colony (chemical communication).

• Food is passed from worker to worker and from worker to the queen and drones, although the food provided for the queen is always royal jelly. Reciprocal feeding continues throughout the life of the bee. A bee up to 2 days old receives more than it transmits. The actual transfer starts by one bee either 'begging' or 'offering' food. Begging bees hold out their proboscis and offering bees fold back their proboscis and open their mandibles exposing a droplet of food.

• During food transfer continual antennal contact takes place; the antennae of both bees are in continual motion touching one another. There is no known antennal language, the purpose is therefore somewhat obscure.

• In addition to the exchange and distribution of pheromones with their chemical messages, the actual food transferred provides information concerning the availability of food and water within the colony. When no nectar is coming in, the bees use stored honey with a high sugar content which prompts water collection. The bee can only metabolise sugars in a 50% solution; ripe honey contains 80% sugar. Conversely, when a heavy nectar flow is available (sugar content generally lower than 50%) the excess water requires to be evaporated and the water carriers are out of business.

• Other information provided by food transfer is:

   • The type of nectar and pollen by taste and smell.
   • Water requirements for cooling; the house bees refuse to receive incoming nectar.
   • Wax building/secretion requirements; if there is nowhere to store the nectar it remains in the crop of the house bees.
   • Older bees (particularly foragers) are usually the ones offering food; this income stimulates queen feeding and increased brood rearing.
   • Indicates to the whole colony the presence or absence of the queen.

• The food transfer process is very rapid throughout the colony. A single transfer between two bees takes about half minute (transferring both food and pheromone). If these two bees feed two others, and then the four feed four others ad infinitum a series 2, 4, 8, 16, 32, 64, c.125, 250, 500, 1k, 2k, 4k, 8k, 16k, 32k, 64k is evolved; ie. 15-16 transfers each lasting about half a minute. In 7 or 8 minutes the whole colony is aware of the 'state of play'. This model is very simplistic and in practice the communication is faster than this. It is readily demonstrated by removing a queen from a colony. In about 5 minutes bees are busily searching the entrance area for her. In a matter of an hour or two emergency queen cells are likely to be started demonstrating the effect of shortage of queen substance.

1.15 An account of the process of swarming and supersedure.

• Swarming and supersedure are survival and reproduction instincts. It is not certain whether swarming developed from what we know now as a mating swarm or whether it evolved from absconding in times of dearth.
• Supersedure is the changing of the queen in a colony without swarming taking place; swarming makes provision for a new queen in the colony if one does not exist before the swarm departs.

• The reasons for swarming and supersedure:

   • An inadequate supply of 'queen substance' to all the workers in the colony is the fundamental reason why this happens.

- With an ageing, damaged or sub-standard queen which can still be laying.
- Due to congestion (overcrowding in the available space) which causes a breakdown in the food transfer mechanism and deprives some of the workers of queen substance although the queen may well be producing adequate quantities.

• When 'queen substance' is in short supply (workers are not receiving the minimum threshold amount), queen cells are produced by the workers and when this happens 3 things can occur:

- swarming can take place,
- the old queen can be superseded,
- the bees will do neither and tear down (destroy) the queen cells they have built.

• It is unknown how the colony decides to proceed. When queen cells are started, the beekeeper must assume that the colony will swarm particularly in the season before the main flow.

• After eggs have been laid in queen cups, the queen is given less food and egg laying falls off. The queen's abdomen contracts and she gets lighter in weight in preparation for the swarming flight. The queen would have difficulty flying when she is in full lay.

• After the sealing of the first queen cell, the first swarm (prime swarm) will issue from the colony depending on the weather and usually between 10·00-15·00 hours. This prime swarm contains the old queen plus 50 - 90% of the colony's bees. If the weather has been continually bad for 8/9 days after sealing of the first q. cell, then one or more virgins may also be included in the swarm.

• A second swarm may issue when the first virgin emerges followed by a further cast or the first virgin out may destroy the others either in their cells or by fighting after they emerge. These 'casts' are very much smaller than the prime swarm. The origin of these casts may be mating swarms and the bees leave with the virgin queen.

• Prime swarms usually emerge and settle close (a few metres) to the old hive and stay for 12 - 48 hours before taking off for their new home. Casts on the otherhand are capricious and seldom cluster for long in any one place.

• Swarming generally occurs in May and June (S. England) but early flora can cause early rapid build up and swarms can occur in April (nb. rape).

• Supersedure generally occurs in August and early September. Often the old and the new queen can be found in the same hive (for a short period), sometimes both laying, until the new queen takes over and the old one rejected. It is not clear whether the new queen kills the old one or whether the bees ball and eject the old one from the hive.

• If the original cause of swarming was insufficient queen substance, the new colony may well be a stable unit ie. less bees for the same amount of queen substance produced by the queen giving each bee the required threshold amount. There is however a strong possibility that this new colony, with the old queen, will supersede later in the season.

• When a swarm settles it has 2 distinct parts:

- an outside shell about 3 bees thick (providing protection and mechanical strength),
- an inner part (loose) consisting of chains of bees connected to the shell.

The outer shell has a distinct entrance to the inside. Dancing (wag-tail) can often be observed on the outside shell indicating the selection of a new home for the swarm.

• The bees in a swarm, before emerging, will gorge themselves with honey ready for wax building at short notice when the new home has been found. For this reason swarms are generally very docile, however care should be exercised with swarms of unknown history.

1.16 How the honeybee passes the winter.

• The essential requirements for the successful wintering of a honeybee colony are:

  • it should have a young queen of good genetic strain,
  • good protection from the elements (sound weatherproof hive in a non- exposed site),
  • adequate stores to survive without feeding (honey and pollen),
  • the colony shall be free of any diseases,
  • good ventilation.

• It is important to note the difference in life expectancy of the summer bee (6 weeks) and the winter bee (6 months). In autumn brood rearing decreases and eventually stops. The young house bees are no longer producing brood food and an enlargement of the hypopharyngeal glands occurs. Similarly, other worker bees consume large amounts of pollen and their glands also enlarge. As a result of consuming this pollen, a considerable amount of fat, protein and glycogen (animal sugar) is stored in the fat body of the bee; these fat bodies increase in number and existing fat bodies increase in size. The fat bodies provide an internal food store which can be used at any time by the bee. Studies undertaken seem to indicate that lifespan is inversely proportional to the amount of brood food produced and the pollen consumed. See the annual colony population graph; the population is steadily decreasing throughout the winter reaching a minimum in about February. There is no abrupt cut off as the bees in the colony change from summer to winter bees.

• The honeybee is poikilothermic (it assumes the temperature of its surroundings ie. the ambient temperature) and the only way it can survive low temperatures is in a cluster:

  • Clustering starts at an ambient temperature of 57°F and at this temperature the cluster is well defined.
  • The cluster has an outer shell (1-3" thick) similar to a clustering swarm. The shell is quite dense with bees entering empty cells.
  • The outer shell temperature is maintained at c.45°F (virtually constant).
  • As the ambient temperature drops below 45°F, the cluster contracts and as the temperature rises above 45°F it expands.
  • Bees in the centre produce heat to maintain the outer temperature at 45°F. The centre temperature ranges from 68° to 86°F when there is no brood present. If brood is present, then the temperature around this brood is maintained at c.95°F.
  • Heat produced = the heat lost. Conduction loss ≈ zero, convection loss ≈ radiation loss.
  • Heat produced by the bees in the centre of the cluster is achieved by a rapid micro movement of the indirect flight muscles in the thorax at the expense of consuming stores.

• It is essential that the cluster maintains contact with the food reserves; without them the colony could perish. Only very occasionally, when the temperature drops rapidly and stays low for a considerable period, the cluster contracts quickly and is cut off from the food reserves.

• Food consumption and respiration produce CO_2 and H_2O; it is essential that both can be disposed of readily. Ventilation is therefore regarded as being extremely important. How this ventilation is catered for in the modern hive is a keenly debated subject amongst beekeepers; it should be remembered that feral colonies do survive very well in some very draughty situations.

• The stores consumed when the bees are clustering tightly when the weather is cold (<45°F) is minimal (a few ozs. per week); it is in the warmer times in winter that a greater food consumption occurs [see appendix 9].

• During these warmer days bees will take cleansing flights and these will also occur on cold days with bright sunshine (very noticeable when snow is covering the ground).

• After the winter solstice and the days begin to lengthen, the queen starts to lay; a tiny patch of brood in the centre of the cluster to start with and gradually expanding. Food consumption starts to rise rapidly, to maintain the incubation temperature. Water is necessary to dilute the stores to 50:50 and if the colony has to forage for water in low temperatures, it can be a very hazardous undertaking for the bees.

- • Other points of interest:
- • Wintering losses are still very high due to poor beekeeping practices.
- • Contrary to popular belief, the bees in the cluster are not continually changing position to keep warm.
- • Note that a connective cluster exists (few hundred bees).
- • If the colony consumes c.30lbs of stores during the winter period, then about 4 gallons of water will be produced and will require to be removed from the hive.
- • Note that honey is hygroscopic and bees uncap stores as required, water being absorbed naturally, thereby diluting them.

** ** ** **

2.0 STRUCTURE AND FUNCTION

2.1 Description of the structure and segmentation of the exoskeleton.

The exoskeleton consists of (see appendix 4):

- epicuticle - thin outside greasy waterproof layer,
- cuticle   - consisting of two parts:
    - (1) exo cuticle - hard sclerotin,
    - (2) endo cuticle - soft chitin,
- epidermis - cellular, which secretes to form the cuticle.
  On the lower side of the epidermis is a basement membrane to which muscles are attached.
  The sensilla of most of the sense organs are formed in the epidermis.

Some of the terms used when describing the exoskeleton and the anatomy of the bee are as follows:

Ventral	under side
Dorsal	upper or back side
Anterior	head end
Posterior	tail end
Lateral	side
Proximal	near to point of attachment
Distal	away from point of attachment

The exoskeleton is divided into three parts:
- head, thorax and abdomen.

2.1.1 Head - is derived from the 6 basic segments (see Dade fig.1 & pl.2)
Appendages - 2 antennae, mouthparts and eyes (5). There has been extensive specialisation of each of the basic segments to form these parts.

Thorax - consists of 4 basic segments: prothorax (T1)
                                          mesothorax (T2)
                                          metathorax (T3)
                                          propodeum (A1)

In many insects there is a constriction between the thorax and the abdomen; in the order hymenoptera this constriction (petiole) has developed between the the first and second abdominal segments of the worm-like ancestors of the insects.

Each true thoracic segment consists of 4 plates:
Tergite   on dorsal surface
Sternite   on ventral surface
2 pleurites on lateral surfaces or the sides.
Appendages - 6 legs (2 on each thoracic segment) and 4 wings connected between tergite and pleurites of T2 & T3

Abdomen - is derived from 9 basic segments consisting of 2 plates only; a tergite and a sternite. Note that there are 10 basic abdominal segments. A1 is on the thorax and A2 to A10 are on the abdomen. On the actual abdomen only 6 can be seen (A2 to A7); A8 to A10 are part of the internal structure and cannot be seen externally.

2.1.2 A large part of the exoskeleton is covered in hair; a vital part of the anatomy enabling the bee to collect pollen. Flexible joints consisting of a membrane with a thin layer of cuticle connect the rigid plates of the exoskeleton. Note the differences in size and shape of the exoskeletons of the queen worker and drone.

2.1.3 Other parts which are visible externally on the exoskeleton are:

- Nasonov gland - located on the dorsal side in the inter segmental joint between A6 and A7 (worker only).
- 10 pairs of spiracles (9 prs. visible; 10th pr. on A8 associated with the sting)
     1st pr. largest on T2, the valve cannot be closed.
     2nd pr. smallest on T3
     3rd pr. largest outlet on A1
     4th to 9th visible on A2 to A7
- 4 prs. of wax glands are situated under sternites A4 to A7 inclusive (worker only). These cannot be seen normally, however the wax scales being produced by worker bees are often in evidence during nectar flows.

2.2 Detailed description of external structure of queen, drone and worker.

Structure common to all three: head, thorax and abdomen:

- Head -   2 compound eyes, 3 simple eyes (oscelli),
               2 antennae (scape + flagellum),
               2 mandibles,
               1 proboscis.

- Thorax -   4 segments (T1 to T3 plus A1),
               2 prs. wings (between tergites and pleurites of T2 & T3),
               3 prs. legs (on segments T1,T2 and T3).

- Abdomen - 6 visible segments (A2 to A7),
               3 invisible segments A8 to A10 (which are internal and part of the sting chamber).

Physical size: Worker about ⅝" long.
               Queen about 1" long (larger than worker in diameter; nb. q. excluder).
               Drone about ¾" long (much fatter than q. or w.).

   Head: Worker   - triangular in shape with long proboscis.
         Queen   - similar to w. but rounder with short proboscis.
         Drone   - almost circular (n.b. large compound eyes).
                    Antenna has extra joint (flagellum 12 segments) and mandibles v. small. The proboscis is short.

   Thorax: Worker/queen - similar in size; dorsal side in q. appears hairless cf. a worker.
           Drone   - larger/stronger, larger wings (stronger flier).

   Abdomen: Queen - very distinctive (long/tapering).
            Drone   - also distinctive (fat and furry).

Worker - specialised (wax and Nasonov glands).
All castes have 10 prs. spiracles on segments T2 to A8, the last being invisible and inside the abdomen.

Legs: All have the same formation - coxa, trochanter, femur, tibia, 5 tarsal joints and a foot (pretarsus).
Note: the fore legs of all 3 castes have an antenna cleaner and only workers have pollen collecting equipment on the rear legs. The hairs on the barsitarsi of the fore legs are used as a brush for cleaning the eyes and head in all castes.

Wings: Forward pr. large with fold to engage with the hamuli on the smaller rear wing. Drone wings much larger. In all 3 castes the wings are folded at rest and lay flat on dorsal side of the abdomen.

Hair: The whole of the exoskeleton is covered in plumose hairs which have an important function in the worker for trapping pollen

2.3 Structure and function of the mouthparts, antennae, legs,wings and sting [It is important to refer to a diagram - appendix 4].

2.3.1 Mouthparts - consist of:

1 clypeus - below the antennae.

1 labrum - hinged to the clypeus and below it. Soft pad (the epipharynx) on the under side of the labrum is shaped to fit the proboscis. When in use it makes an airtight seal around the proboscis at the laciniae to allow it to function as a sucking pump.

2 mandibles - working in horizontal plane are hinged to the genae. Concave and ridged on inner side. Both have grooves which connect to a duct in the mandibular glands. Note the differences in q., d. and w. mandibles.

Function: shovelling food into the mouth, handling, biting, cutting and kneading wax, building comb, collecting and applying propolis, feeding brood food and pollen to larvae, dragging debris out of hive, grooming, fighting and for supporting the proboscis when extended.

1 proboscis - (complicated structure see diagram in Dade)
It consists of 2 basic parts - the central portion (glossa,etc.), - outer portion (maxillae, labial palps).

• Centre - postmentum, prementum, paraglossa, glossa and flabellum. The last 3 are surrounded by 2 labial palps.

• Outer - 2 maxillae, each consisting of stipe, lacinia and galea.
• The maxillae are connected laterally to the postmentum by two lorums and by two cardines to the fossa (in a fore and aft direction).
• The labial palps and the galea are formed into a hollow section with the glossa in the middle. Honey, nectar and water are sucked up this section via the inter-space (food canal) outside the glossa.

• Saliva (mixed secretions from the postcerebral and thoracic glands) flows into a pouch (the salivarium) near the junction of the glossa and the prementum. It runs down inside the glossa ('C' shaped section) and mixes with nectar or syrup taken up the food canal.
• The proboscis when at rest is folded underneath and is hinged at the junction of the prementum and paraglossa.
• The proboscis when in use is extended forward and raised to mouth level, sealed by the epipharynx of the labrum and grasped by the mandibles to steady it. Honey, nectar or water are then sucked up.
• If the quantity of liquid is small, the end of the glossa (hair covered) picks up small droplets due to surface tension and the glossa is then retracted into the food canal where it can then be sucked up as before.
• Note the difference in length of the proboscis in the three castes and the variations that occur between different species of apis mellifera.

2.3.2 Antennae.

The bee has two antennae each consisting of:

> • the scape (nearest to the head),
> • the flagellum (segmented, 11 segs. in q. and w., 12 in d.).
>   > Note 1st segment is called the pedicel.

The antennae are the sense organs used for a variety of sensing purposes. The flagellum of the antenna is covered in sensilla (sensors connected to nerves through the exoskeleton) of different types and used for different purposes as follows:-

• touch	sensilla trichodia	hairs
• smell	s. basiconica	pegs
• CO_2, RH, T	s. coeloconica	pits
• stress/strain	s. campaniformia	bells
• smell	s. placodea	plates

Sensilla appear in great numbers on the 8 distal segments. The s. placodea associated with the 3 castes are as follows:-

• Queen	-	2 - 3000
• Worker	-	5 - 6000
• Drone	-	c. 30,000

A drone can detect the presence of a queen in flight at a distance of c. 50 metres using the s. placodea to detect the pheromone "queen substance". Experiments amputating the segments on the flagellum show that the senses are virtually lost when 8 segments are removed.

The 'Organ of Johnston' is situated in the pedicel and is believed to detect vibrations and is used by the bee as a wind speed indicator.

The antennae are used continually during food transfer and appear to be play an important part in this function. However, no antennal language has yet been discovered.

2.3.3 Wings.

- The wings are used for:
    - Flying and hovering (forwards, backwards, up and down).
    - Fanning for - ventilation (cooling),
        - ripening honey,
        - distributing pheromones (e.g. Nasonov).
- 2 pairs wings (fore and rear; fore wing on segment T2 and rear on T3. Fore wing is the larger of the two.
- When wings are furled, and at rest, they are separate one from the other. As they are unfurled, ready for flight, they are joined by hamuli (hooks) on fore edge of the rear wing connecting on to the fold on the rear edge of the fore wing. There are about 20 hamuli on each of the rear wings, the exact number can vary slightly.
- The wing roots attach to tergite and pleurite on segments T2 & T3 - pleurite provides a fulcrum for movement of the wings.
- In flight, the wings are operated by large indirect muscles in the thorax; these muscles drive only the fore wing, the rear wing trails on the fore wing via the coupling at the fold and hamuli. These indirect muscles only provide up and down movement to the wings.
- There are direct muscles attached to all 4 wings, 4 muscles on each fore wing and 3 on the rear wings. These direct muscles are used only for furling/unfurling and for 'trimming' (i.e. twisting) the wings to correct for yawing, rolling and pitching during flight.
- Vibrations while in flight = c. 200/250 c.p.s.
- Speed of flight = c. 15 m.p.h. but can increase to 25 m.p.h. for short periods.
- Blood sugar content for flight = c. 2% (1% cannot fly, 0·5% cannot move).
- Rate of fuel (sugar) consumption = 10 mg/hr during flight; this is c. 50 times greater than at rest.
- Range = 4/5 miles (c. 15 mins.); can be extended by resting while glycogen stored in body is converted to sugar.
- Note the use of the scutal fissure for operating the wings during flight.
- The venation in the wings is used as a biometric aid in the identification of various strains of bee (taxonomy).

2.3.4 Legs.

The bee has 3 pairs of legs; appendages to T1, T2 & T3. They are used for: standing, walking and clinging to a variety of surfaces at any angle or upside down. The most important use, however, is for cleaning, comb building and the collection and transportation of pollen and propolis.

- Each leg has 6 major divisions which can be articulated separately by internal muscles; these are:

    - coxa
    - trochanter
    - femur
    - tibia
    - tarsus (divided into 5 tarsomeres)
    - pretarsus (consists of 2 claws for rough surfaces and an arolium for smooth surfaces) or foot.

- Forelegs: are small and close behind the head (ALL CASTES). Hairs on the 1st tarsomere and basitarsus are used for cleaning dust/pollen, etc. from the head. Circular notch in basitarsus and spur (fibula) on the tibia form the antenna cleaner. The circular notch is lined with fine hairs to provide a comb.

• Middle legs: Single spine on tibia has no known use. This leg has no specific function except for the pollen brushes on the inner surface of basitarsus for cleaning pollen from the thorax and passing it to the rear legs.

• Rear legs: Pollen press and basket (WORKER BEE ONLY). The harvest of pollen collects on the pollen brushes of the basitarsus. While hovering, the bee rubs her hind legs together and the pollen is raked by the rastellum (pollen brushes) of the opposite leg on the distal end of the tibia. The pollen falls into the auricle and the tibio-tarsal joint is closed and pollen is squeezed and forced to emerge onto the outer side of the leg to be caught in the long hairs of the corbicula on the tibia. Back in the hive, the pellets of pollen are disengaged by the middle legs and dropped into cells. Propolis is bitten off with the mandibles and also carried back to the hive in the corbicula but unloaded by another bee. Wax plates are removed from the mirrors of the abdominal sternites by the rear legs. The drone and the queen have no specialised functions for the use of the rear legs.

2.3.5 The sting.

• The actual sting consists basically of 3 parts: stylet and 2 lancets (barbed). There are other associated parts namely, bulb and umbrella valves, the rami, oblong plate (fixed), quadrate plate (moveable), triangular plate (moveable), muscles connected to the oblong and quadrate plates, acid sac, acid gland and alkaline gland.

• The lancets slide on tracks on the stylet and the 3 parts form a tube (the venom canal).

• The stylet and lancets are connected to a bulb via vents and umbrella valves which deliver the venom from the bulb down through the venom canal.

• Each lancet is connected to a corresponding ramus which in turn is connected to a pivoting triangular plate. Movement of the triangular plate is rotational, activated by a moving quadrilateral plate and also a fixed oblong plate. One corner of the triangular plate is pivoted to the fixed oblong plate..

• As one lancet is pushed forward, a simultaneous action withdraws the other and vice versa.

• Movement of the ramus and lancet also operates the umbrella valve in the bulb, thereby ejecting venom from the bulb into the venom canal.

• The bulb (venom reservoir) is connected to the venom gland via the acid sac. Another gland, the alkaline gland appears to supply the bulb but actually secretes into the sting chamber.

• The use of the alkaline gland is not exactly known. Suggested uses are lubricant for the stylet and in the case of the queen for gumming eggs to the base of the cell. Neither uses are proven. The acid gland produces the venom which contains:

  • Phospholipase A *   enzyme
  • Hyaluromidase *   enzyme
  • Acid phosphatase   enzyme
  • Allergen C *
  • Mellitin *
  • Mellitin F
  • Mast cell degranulators (peptide. secapin, tertiapin)

* = main allergens.

Note that the above are the main constituents of bee venom, but other substances are present in small quantities.

• Use of the sting - only known to be used for self defence and defence of the colony. It should be noted that different strains of bee have a greater or lesser degree of colony defensive behaviour.

• When a bee leaves its sting in the victim and tears itself away, the associated ganglion and muscles are still attached to the sting and the lancets continue to operate, the sting penetrating deeper and deeper. It should be scraped out with a knife or finger nail as quickly as possible. The enzymes in the venom cause the release of hystamine from mast cells which causes swelling. Other allergens effect the nervous system.

• The queen only uses her slightly curved sting against rival queens. Most books state that her sting is barbless; there are however 2 or 3 small barbs at the tip of the sting.

2.4 Description of the general structure and function in the worker bee.

2.4.1 The alimentary canal and the digestion of sugar and pollen.

• The alimentary canal (transmitting food through an animal body from mouth to anus) consists of :

- • pharynx (mouth),
- • oesophagus (tube),
- • honey sac or honey stomach or crop,
- • proventriculus (valve),
- • ventriculus (mid gut),
- • pyloric valve,
- • small intestine (+Malpighian tubules),
- • rectum,
- • anus,
- • salivary glands - postcerebral and thoracic.

Note that the salivary glands are not strictly part of the alimentary canal (see definition above) but are often included in literature on the bee and its anatomy. See section 2.4.4.

• Pharynx: the true mouth cavity with:

- • inlets from hypopharyngeal glands and proboscis,
- • outlet to oesophagus,
- • inlet/outlet via the mandibles.

• Oesophagus: tube through the thorax connecting pharynx to the honey sac. Its only function is to provide a passage for food in both directions.

• Honey sac: transparent bag at the anterior end of the abdomen and capable of considerable dilation. Max. load c. 100mg./average load 20 to 30mg.

- Proventriculus: is a one way valve which:-

  - prevents (when necessary) nectar and honey flowing into the ventriculus.
  - separates the pollen from the nectar/honey. The nectar is retained in the honey sac and the pollen is collected in pouches behind the lips of the valve; the filtration being done by very fine hairs [capable of filtering to 1μ]. When a pouch is full, a bolus of pollen is passed into the ventriculus.

- Ventriculus: is the true stomach of the bee where digestion of foods take place. Note the shape which is like 'gas mask tubing' which is ideally designed for peristalsis which moves the food and waste matter through the final parts of the alimentary canal. The ventriculus is lined with an epithelium which is proliferating cells continually; these sloughed off cells contain enzymes. The enzymes enter the pollen grains through the germ pores and digest the proteins in the grain. The resulting nutrients are then absorbed through the walls of the ventriculus and into the haemolymph (blood) in the abdominal cavity. A similar digestion process occurs with honey and nectar, the enzymes breaking down the polysaccharides into monosaccharides which can then be absorbed into the haemolymph through the walls of the ventriculus. The epithelium is covered with a jelly like substance, the inner surface of which forms the peritrophic membrane. This encloses the food and is said to prevent abrasion of the ventricular wall. Waste matter produced in ventriculus contains pollen husks, fat globules and uric acid and nitrogenous waste from the Malpighian tubules which enter the alimentary canal at the posterior end of the ventriculus.

- Pyloric valve: this is a thickening of the walls at the anterior end of the small intestine just behind the entry point of the Malpighian tubules. The valve is lined with backward facing hairs said to assist in directing the contents in the backward direction.

- Small intestine: is constructed in a fluted formation of 6 flutes. This type of formation provides a large surface area and slows down the passage of food. This suggests that absorption of digested food could take place in this region.

- Rectum: this is similar to the honey sac in so far as it can be dilated to such an extent that it can fill the whole abdomen. Six rectal pads are found on the outer surface of the rectum. They are quite distinctive but the function is so far unknown. It has been suggested that they extract water from the contents and return it to the haemolymph; the microscopic structure does not support this. Other suggestions are for the absorption of digested fats and maintaining the concentration of salt in the blood.

- Anus: the final outlet from the alimentary canal.

- Malpighian tubules: there are about 100 of these tubules, closed at the distal end and joining the alimentary canal adjacent to the pyloric valve. They spread throughout the abdominal cavity and absorb through their walls nitrogenous waste from the blood.

- Note that 2 adult bee diseases are associated with the alimentary canal, these are:

  - Nosema apis (Zander) - in the ventriculus restricting the digestion of proteins from the pollen grains.
  - Malpighamoeba mellificae (Prell) - in the Malpighian tubules.

2.4.2 Excretory system including function of the Malpighian tubules.

• The excretory system consists of:

  • the Malpighian tubules,
  • the small intestine,
  • the rectum and anus,
  • the tracheae.

The tracheae discharge both CO_2 and water vapour from the spiracles and therefore are technically part of the excretory system although many text books do not treat them as such.

• The Malpighian tubules.

  • Approx. 100 Malpighian tubules (named after the Italian biologist/entomologist, Malpighi) enter the posterior end of the ventriculus adjacent to the pyloric valve. They spread throughout the whole of the abdomen, are surrounded by blood and are closed at their distal ends. They are the principal organs of excretion.
  • The walls of the tubules are composed of a single layer of cells and through their walls nitrogenous waste and salts are absorbed from the blood. This waste passes down into the ventriculus and then into the small intestine.
  • It is in these tubules that the cysts of the protozoan parasite Malpighamoeba mellificae are found. Curiously these cysts seem to have no ill effect on the bee, although it is difficult to comprehend how the tubules work effectively in severe infections when the tubules are 'solid' with cysts.

• The small intestine.

  • Entry is from the ventriculus via the pyloric valve which is itself formed by a thickening of the walls of the small intestine at this point. the small intestine is constructed in a fluted configuration, the 6 flutes running longitudinally providing a large surface area. The inside of the pyloric valve is small in diameter (cf the ventriculus and small intestine) and contains fine backward facing hairs which prevent digested food from returning into the ventriculus.
  • The constriction of the pyloric valve coupled with the large surface area of the small intestine is compatible with slowing down the passage of food and absorption of it in the small intestine. Excess waste water is absorbed through the walls of the small intestine and passed with other waste products to the entrance of the rectum at the posterior end.

• The rectum.

  • The rectum is a flexible transparent bag with 6 rectal pads which are quite distinctive and are compatible with the fluting on the small intestine. The use of the rectal pads is unknown.
  • The rectum stores the faeces until the bee can exit the hive and fly. The bee normally defaecates on the wing. In winter when the bee cannot fly outside the hive, the rectum is capable of distending to fill a very large part of the abdomen. Any fouling of comb or top bars during winter means dysentery and if Nosema is present, it is spread around the colony by food transfer as the bees try to clean up the fouled combs. Excess water, causing dysentery, cannot be retained in the rectum and is the basic cause of comb fouling in winter.

• Faeces.

    • Faeces are evacuated via the anus and are made up of:
        - indigestible starch (dextrine),
        - pollen fat (globules),
        - pollen husks,
        - exhausted epithelial cells from ventriculus,
        - nitrogenous waste (eg. uric acid) and salts,
        - water.

• Tracheae.

    • The tracheae form the main 'highways' of the respiratory system and provide an inlet passage for O_2 and an outlet passage for CO_2. See section 2.4.3

2.4.3 The respiratory system (interchange of O_2 and CO_2).

• The respiratory system consists of:

    • spiracles (vents at the sides of the thorax and abdomen),
    • tracheae (walled tubules),
    • tracheal sacs (expandable bags),
    • tracheoles (fine tubules where O_2 is absorbed and CO_2 exchanged).

• Spiracles: 10 pairs, 3 on thorax, 7 on abdomen (6 visible, last near sting is invisible)

    • 1st - on T2 of thorax, it is the largest and cannot be closed, (nb. Acarine). It provides entry for the major air intake during flight.
    • 2nd - on T3 of thorax, it is the smallest and concealed.
    • 3rd - on A1 of thorax, it is also large and the major outlet for CO_2 during flight.
    • 4th to 9th - on A2 to A7 of the abdomen.
    • 10th - cannot be seen externally as it is contained in the sting chamber.
    • All the spiracles have valves except T3. the valve on T2 cannot be closed although it does have muscles apparently for this purpose. The valves on A1 to A8 are identical, the spiracle being the entry into a chamber (atrium) with the valve at the inlet to the trachea.
    • Larvae have the same number of spiracles but only use one side from the time they hatch until sealing because they are lying in a pool of brood food and spiracles on the lower side are below the surface of the liquid food.

• Trachea: connect the spiracles to the tracheal sacs. They are maintained in the dilated state by spiral thickenings (taenidiae) in the cuticle wall. 1st spiracle and trachea on T2 is branched to both the thorax and the head. All other trachea lead to the tracheal sacs.

• Tracheal sacs: Thin walled flexible air sacs which are expanded by air and contracted by blood pressure as a result of muscular action causing a 'bellows type' movement of the abdomen, nb. breathing (inlet of O_2, outlet of CO_2). There are 2 large sacs in the abdomen as well as smaller ones in both the abdomen, thorax, head and even the legs. The purpose of these sacs is to act as a reservoir when air is being used rapidly, the trachea and tracheoles, being small in diameter, would create a too greater resistance to flow for the amount of air being used.

• Tracheoles: These are the branching tubules from the tracheal sacs which spread out to all parts of the bees anatomy. The cavities of the exoskeleton are literally full of them. They are surrounded by blood and the open ends also contain some blood. This network of tracheoles is in close proximity to all the tissues (eg. muscles). Oxygen from the air in the tracheoles diffuses into the blood and carbon dioxide, the result of the oxidisation process of metabolism, is removed via the same route.

• The major use of O_2 is while the bee is in flight, in order to supply the large indirect muscles in the thorax. These muscles when dissected are a bright pink colour due to their cytochrome content, a substance which facilitates the gas exchange.

• Other points: CO_2 is used as an anaesthetic for queen bees during artificial insemination. Insects can withstand much higher concentrations of this gas than homo sapiens. The normal concentration in the brood nest $\approx 1\%$ and if this rises $\approx 2\%$ Chalk Brood spores will germinate, a necessary condition for their reproduction.

2.4.4 The exocrine glands.

• The exocrine glands (those secreting externally) are:

  • hypopharyngeal,
  • mandibular,
  • Nasonov,
  • wax,
  • thoracic and postcerebral (salivary glands).
[Note that some books correctly regard the sting as an exocrine gland because it has glands which secrete externally. The secretions are the alarm pheromone, iso pentyl acetate, as well as the venom].

• Hypopharyngeal: located above the pharynx and under the frons. There are 2 glands one on each side of the head. An axial duct from each gland opens separately via 2 small pores on to the hypopharyngeal plate and the secretion accumulates in the food channel in the base of the proboscis inside the pharynx (mouth).
  • The secretion from the glands is called brood food or royal jelly. They are basically the same but there are chemical differences depending on whether it is being fed to worker or queen larvae in order to produce the two female castes. It should be noted that the adult queen lives solely on a diet of royal jelly which is very high in protein. The feeder bee folds back its proboscis, opens its mandibles and raises its labrum to allow the entry of the queen bee's proboscis.
  • The glands are large in the worker, rudimentary in the queen and non existent in the drone. Glands are thread like with round cells (acini) which are active in workers 5-15 days old (nurse bees) after which the acini shrink and become inactive (they become atrophied). It should be noted that older foraging bees can, if the needs of the colony so require it, consume pollen and re-activate their hypopharyngeal glands for brood feeding. Workers deposit the brood food in the cells, directing it into place with their mandibles.
  • In older bees the hypopharyngeal glands are believed to be the source of invertase although extracts from the glands have not been separated for analysis. Dade makes the point that analysis of the secretion of the postcerebral glands show surprisingly differing results, some workers indicating the presence of invertase others denying it. For practical purposes modern thinking is 'invertase from the hypopharyngeal glands'.
  • Brood food (bee milk in some books) and royal jelly contains mainly:
    - proteins,

- several vitamins of the B group,
- vitamins C & D but not E,
- (E)-10-hydroxy-2-decenoic acid [10-HDA], (which acts as a preservative preventing bacterial infection of the food while it is laying in the cell) but note this comes from the mandibular glands.

• Mandibular: sited in the head under the genae and immediately above the mandibles. Ducts open into grooves in the mandibles, the outlets being controlled by muscles.

  • The glands are very large in the queen, smaller in the worker and rudimentary in the drone.
  • Secretions contain:
      Worker - 10-HDA (preservative),
            - 2-heptanone (alarm pheromone),
            - ????????? (for working wax).
      Queen - (E)-9-oxo-2-decenoic acid [9-ODA],
            - (E)-9-hydroxy-2-decenoic acid [9-HDA],
            - ????????? (other unidentified pheromones),
            - 13 other identified pheromones.

• The secretion from the queen glands is popularly known as queen substance; it is very important in controlling colony behaviour. It has two known functions inside the hive, namely:

    - inhibiting queen cell production
    - inhibiting worker ovary development

It is distributed around the colony by food transfer and is very attractive to worker bees. Outside the hive queen substance is a drone attractant during queen mating flights and also holds swarms together as a cohesive unit.

• Nasonov: scent gland located at the anterior end of tergite A7 in a transverse groove adjacent to the posterior end of tergite A6. It is exposed in use by flexing the two abdominal tergites to allow the secretion to evaporate; the process being assisted by fanning.

  • The gland secretes into the scent canal (transverse groove) the secretion containing:
            - E citral and geranic acid, the 2 most important components,
            - Z citral, geraniol, nerol, nerolic acid and E,E farnesol.
  • Maximum production is found in foragers; it is low in winter and high in the spring.
  • The pheromone is highly attractive to bees and is in much evidence when bees are shaken, swarms being hived, etc. It is only found in workers and is used generally at or near the hive entrance. Colloquially known as "the come in and join us" pheromone.
  • Associated with the collection of water (foragers release Nasonov pheromone) but not with nectar or pollen. Also used on rich sources of sugar syrup as this also is odourless.

• Wax glands and wax production [see appendix 4]: the glands are located inside the exoskeleton on sternites A4-A7 inclusive. There are 2 glands on each sternite, making 4 pairs in all. The glands secrete a liquid which passes through the mirrors and oxidises as a flake of wax in the wax pockets. The glands, mirrors and pockets being known colloquially as the "waistcoat pockets".
• Wax is secreted at relatively high temperatures (33°-36°C) after consumption of large amounts of honey. Various estimates are quoted for the amount of honey to metabolise 1lb of wax. 5-8 lbs. being a realistic estimate.

• Wax glands are best developed in worker bees 12-18 days old.
• When building comb, bees hang in festoons near the building place, after gorging themselves with honey, waiting for the wax to form. See section 1.8.6 for details on comb building.
• The wax glands inside the exoskeleton are covered with fat bodies and other cells. The major components of beeswax are:

- hydrocarbons	16%
- monohydric alchohols	31%
- hydroxy acids	13%
- fatty acids	31%
- diols	3%
- other substances	6%

The chemistry of how the wax is produced and how it diffuses through the mirrors is extremely complex and it is not necessary to know the detail; however it is necessary to know that a diffusion process is involved.
> • Wax is normally white but can be tinged with yellow hues caused by pigments that originate in pollen (eg. when a colony is working dandelion, new comb is noticeably coloured yellow).
> • For completeness a few of the physical properties of wax are:

> - SG = 0·95 [Honey like odour & faint taste],
> - melts at 147·9°F, solidifies at 146·3°F. [when pure]

• Salivary glands (thoracic and postcerebral): there are two pairs of glands, one behind the brain (postcerebral) and the other in the thorax. From these glands a duct from each emerges. The two ducts join, in the head, into a common duct leading to the salivary syringe at the top of the proboscis. Secretions from both glands run down the glossal tube and mingle with food which is then sucked up the food canal of the proboscis.
> • The secretions are used to dilute honey and to dissolve sugar crystals at times when water is scarce. The secretions are slightly alkaline and their use is not fully understood.
> • Analysis of the secretions have given very diverse results.
> • The thoracic glands in the adult worker bee are derived from the silk glands of the larva and not from the thoracic glands (more correctly pro-thoracic glands) of the larva which are used for secreting hormones during its development stage.

2.4.5 The circulatory system (heart, dorsal and ventral diaphragms).

• The circulatory system consists of:

> • heart,
> • dorsal diaphragm,
> • ventral diaphragm,
> • blood (haemolymph),
> • aorta,
> • antenna vesicle.   [see system diagram, appendix 4]

• Heart: elongated organ of 5 segments laying just under and extending along virtually the whole length of the roof of the abdomen. It has 5 pairs of openings or one way valves (ostia) allowing the blood to enter the heart when it is dilated. Conversely the ostia are closed when the heart is contracted. It is closed at the posterior end and the anterior end leads directly into the aorta.

- Heart is suspended from the dorsal side of the exoskeleton and is also attached to points on the dorsal diaphragm by branching threads at the ends of the muscles in the dorsal diaphragm.
- Heart beats forwards from behind in successive waves of contractions produced by the muscles. It is not proved that the muscular action of the heart is controlled by the nervous system. No nerve connections have been found to the appropriate muscles.
- Function of the heart is to circulate the blood around the whole body of the bee.

• Dorsal/ventral diaphragms: are thin transparent membranes which are attached at points on tergites and sternites. From the attachment points (dorsal A3-A7, ventral T2- A7) muscles fibres radiate.

- The diaphragms are responsible for setting up a circulation inside the abdomen and drawing blood into the abdomen from the thorax.
- By contractions of the diaphragm muscles, waves are set up in the thin sheets which, in turn, create currents of blood between them and the abdominal walls of the exoskeleton.
- The blood flow is forwards in the dorsal sinus and backwards in the ventral sinus.

• Aorta: is continuous with the heart and passes into the thorax at the petiole. At this point it is convoluted and then passes between the indirect flight muscles in the thorax as a simple tube and then on to the head where it is open ended just behind the brain.

- The convoluted tube acts as a heat exchanger extracting heat from the thorax muscles cooling them as blood flows inside the aorta in the forward direction.

• Antenna vesicle: without this vesicle there would be no way of supplying blood to the antennae. This pulsating vesicle has two tubular outlets which pumps blood to the ends of the antennae. The tubes run down the antennae parallel to the antennal nerve. The return flow is outside the vesicle tubes.

• Blood: the whole of the body, heart, etc. is filled with blood together with legs and wing roots which are contiguous with the body shell. Blood is colourless plasma and contains haemocytes (white blood corpuscles) which act as phagocytes by ingesting and destroying bacteria. The flow of blood is generated by the heart and diaphragms, the pumping action creating a pressure gradient, high at the head and low in the abdomen, which results in a return flow.

- The function of the blood is as follows:
    - to carry nutritive substances from the alimentary tract to cells and tissues where they are consumed.
    - to transport the waste products of metabolism from tissues to the Malpighian tubules,
    - O_2 diffuses through the blood at the end of the tracheoles, these tracheoles transport the gas to the tissues for respiration,
    - similarly CO_2 diffuses through the blood in the reverse direction to be eliminated via the spiracles,
    - distributes heat around the body. Note that the bee is poikilothermic and takes up the temperature of its surroundings.
    - to fill the complete body cavity, maintaining it under pressure and thereby maintaining the body shape [turgor effect].

** ** ** **

3.0 FORAGING, NECTAR and HONEY

3.1 Composition of nectar and its variations.

• Nectar is composed of:

- water,
- sugars (5 - 60%), typically 20 - 40%,
- other substances - salts, acids, enzymes, proteins and aromatic substances.

• The sugars are principally sucrose(s), fructose (fr) and glucose (gl). The nectar types are discrete to each plant species which are generally in three categories, namely:

- sucrose dominant (eg. long tubed flowers such as clover, etc.),
- balanced nectar (roughly equal amounts of s, fr and gl),
- fr or gl dominant (eg. rape which is gl dominant like most crucifers).

• Bees tend to prefer balanced nectars but the nectar with the highest overall sugar content is usually collected in preference to one of lower sugar percentage.

• Gl/fr ratio of nectar - high, granulates quickly with a fine grain - low, granulates slowly with a coarse grain. nb. glucose is less soluble in water cf. fructose or sucrose.

• Variations in nectar:
- composition by flower species - each is different,
- composition depends on:
- weather conditions (temperature, humidity, wind speed, sunlight),
- soil conditions (water content, PH, type such as chalk or clay, etc.).

The variations in nectar secretion of plants is extremely complex with a large number of variables. Sunlight is very important as this is necessary for photosynthesis (hydrocarbons to nectar). With plenty of sunlight temperatures increase and a minimum T is required for the enzymes causing nectar secretion to operate. Rain may wash the nectar from the flower or dilute it or wind may dry the nectar evaporating some of the water and increase its sugar content; both conditions can sometimes be found on the same tree.

• Variations in the amount of nectar available in a given area is important to the beekeeper; only a finite quantity is available. An area can be over stocked with bees to the detriment of the beekeepers concerned.

• A few figures on foraging areas and colony density:

- Dr. Bailey considers that colony density should be no higher than 1 colony per 10 sq. kilometres to minimise disease (or 1 colony/4 sq. miles approx.).
- The foraging area of 1 colony = $\pi 3^2$ = 28·3 sq. miles [r=3 miles].
- ∴ 7 colonies/apiary is about the maximum.
- cf. 1 colony/ acre for pollination purposes or on a concentrated nectar crop such as rape. Note that 1 sq. mile ≈ 640 acres.

• It is important that the beekeeper should get to know his area; the flora, the micro climates, other beekeepers and where each keeps his bees all in relation to available nectar supplies.

3.2 The way nectar is collected and conveyed back to the hive.

• The factors which encourage nectar collection have, surprisingly, been little studied. It is not known whether the amount of nectar collected is related to the amount of honey stored; it is obvious that collection extends far beyond the colony's actual requirements.

• The presence of a queen and brood stimulates the collection of nectar in much the same way as it stimulates pollen collection. Unlike pollen which is deposited directly by the forager into a cell, nectar foragers pass their load to a house bee and can also do a wag-tail dance to recruit more foragers.

• Scouting and finding the source of forage occurs first. About 2% of the bees in a colony actually scout for forage. The scouts return with a load and communicate the source by dancing. The forage selected by the colony is likely to be the best in quality (highest sugar content) and quantity.

• Returning foragers are likely to repeat the wag-tail dances. The number of foragers in a balanced colony is about one third of the population (ie. two thirds of the adults are house bees).

• The bee collects nectar by sucking up the food canal of the proboscis (see section 2.3.1) and thence to the honey sac in the abdomen via the pharynx and oesophagus. Average load is 40mg. (cf. the weight of bee = 90mg.) taking approx. 100 - 1000 visits to flowers on the one foraging trip. Foragers make c. 10 trips per day ranging in time from 30 - 60 minutes.

• The enzyme invertase from the hypopharyngeal glands is added to the nectar as it transits the pharynx to the oesophagus and the conversion of sucrose (disaccharide) to fructose and glucose (monosaccharides) starts on the flight back to the hive. The process is continued after reception by the house bees receiving the load.

3.3 Conversion of nectar to honey including chemical changes and storage of the honey by the bee.

• The conversion of nectar to honey involves two changes:

  • chemical change (disaccharide to monosaccharides),
  • physical change (evaporation of water).

• Chemical change: A forager returning to the hive with a load of nectar transfers the load to a house bee which then undertakes the completion of the chemical change which was started by the forager as follows:

  • A small droplet of nectar is re-gurgitated into the fold of the partly extended proboscis and then swallowed (time about 10 seconds). This process is repeated 80 - 90 times in about 20 minutes on the same droplet which is then deposited in an empty cell or ½ full cell. This re-gurgitation process = 'ripening'.
  • Sucrose is converted to glucose and fructose by the enzyme invertase from the hypopharyngeal glands which is further added by the house bee during the ripening

process. Note that the nectar will contain all three types of sugar in varying quantities depending on the floral source (see section 3.1).
- During the ripening process, the water content is reduced by approx. 15% as a result of evaporation when the droplet is exposed.
- Finally the house bee undertaking the ripening hangs the unripe honey (now honey not nectar) to dry in either empty cells or ½ filled cells.

• Physical change: is the process of evaporating the excess water in the unripe honey to bring the sugar concentration up to about 80%. This is done as follows:

- A large amount of space is required [see section 6.14] as the honey is hung in empty or partially filled cells in order to provide the maximum surface area for evaporation purposes. In the empty cells, the honey is deposited with a 'painting action' on the upper surface of the cell.
- Currents of air are distributed around the hive by the bees fanning bringing in dry air and expelling moist warm air (see sections 1.8.7 and 1.8.8).
- As the water content diminishes and the sugar concentration of the honey approaches 80%, the honey is moved and the partially filled cells are completely filled and capped with a pure wax capping with a minute air gap beneath the capping.

• Other points for consideration:

- It will be clear from the above that plenty of space is required and there is much sense in the old adage 'over super early in the season and under super late in the season'.
- It is important to provide conditions in the hive to allow the bees to ventilate and ripen their honey easily. The authors believe that by providing top ventilation it assists the bees to ventilate via the hole in the crown board and roof ventilators. In a nectar flow if the roof is raised there are always bees fanning around the open feed hole; we notice many beekeepers keep this hole closed for no apparent reason. It must be hard on the bees to move the air from the 3rd or 4th super down to the bottom entrance.

3.4 Constituents by percentage of an average honey.

• The composition of an average honey has the following approx. values

- 18% water,
- 35% glucose (dextrose),
- 40% fructose (levulose),
- 4% other sugars (eg.maltose, raffinose, etc.),
- 3% other substances:
    about 15 organic acids (eg. acetic, butyric, malic, etc.),
    -about 12 elements (eg. potassium, calcium, sulphur, etc.),
    -about 17 amino acids (eg. proline, glutamic, lysine, etc.),
    -about 4-7 proteins (0·2%) [Note, in heather honey the protein value is very high c.1·8%].

• The 3% other substances are responsible for providing each type of honey with its own characteristics eg. colour, aroma and taste.

• Generally, less colour (lighter) = less flavour.
• It should be noted that honeydew has a different composition to the average honey quoted above.

3.5 Description of the granulation process of honey.

• Honey granulates because it is a supersaturated solution of more sugars than can normally remain in solution. Such solutions are more or less unstable and, in time, will return to the stable saturated condition.

• Glucose is less soluble in water than fructose. In many honeys the glucose content is in an unstable condition and therefore they granulate (crystallise) readily (eg. rape and other brassicas).

• The ratio glucose/fructose has been widely used to determine which honeys granulate readily (a high ratio indicating rapid granulation). However, the glucose/water ratio is more closely related to granulation tendency;

> • 1·7 or lower, no granulation occurs.
> • 2·1 or higher, predicts rapid granulation.

• There is an optimum temperature 57°F (13-15°C) for granulation to occur. Higher and lower temperatures reduce the rate of granulation. As the temperature reduces the viscosity increases slowing down any molecular movement and as the temperature increases molecular activity increases and any crystals tend to melt.

• High viscosity honeys (dark) promote slow granulation with large crystal. Low viscosity honeys (light) have rapid granulation with fine crystals.

• For granulation to occur a seed must be available:

> • say 5% of a granulated honey (will emanate seed crystal),
> • dust, pollen grain, bubble, etc.

Seed is likely to occur more readily in large quantity containers (eg. 30 lb buckets) rather than in small 1 lb jars.

• Flavours and bouquets are mostly lost by granulation - the same happens with heating [nb. volatile oils etc. evaporate].

• The enzyme diastase in honey breaks down starch and will reduce with time or by heating. Activity is measured by 'The Diastase Number' which is measurable to assess age and heating of a sample. Similarly, the hydroxymethylfurfuraldehyde (HMF) level can be measured which is also a measure of age and the amount of heat that has been applied to a sample. As honey is heated or ages the diastase number decreases and the HMF value increases. Both levels are tested by the 'Weights and Measures' when examining a random sample offered for sale.

3.6 Description of the process of fermentation in honey.

• Fermentation is caused by sugar tolerant (osmophilic) yeasts reacting with glucose and/or fructose producing alcohol and CO_2. Note that alcohol + O_2 → acetic acid + water with an accompanying characteristic smell. Although these yeasts are known as sugar tolerant (cf. say wine making yeasts) it is generally necessary for the water content of the honey to increase by a few percent ie. from 17% to 19 or 20% in order for fermentation to start.

• The source of these 'wild' yeasts are flowers and in the soil. Generally honey contains these yeasts in greater or lesser numbers.

• Fermentation is more likely when:

   • the honey is unripe before extraction.
   • the jars are not air tight (ingress of moisture - honey is hygroscopic),
   • the honey granulates (higher water content between crystals).

• Generally, yeasts are only active between 50°F and 85°F. Ideally to prevent any fermentation honey should be stored outside these temperatures.

   • Storage below 50°F discourages granulation and fermentation.
   • Storage above 80°F damages honey (nb. HMF and diastase values) and should therefore be avoided.

• Most yeasts are killed at temperatures above 120°F. Therefore, the heat treatment of run or liquid honey at 140°F for ½ hour will not only prevent incipient granulation but will also ensure no fermentation.

• Moisture content is very important:

   <17·1% water unlikely to ferment,
   17·1 - 18% with yeast count 1000/gm - OK for 1 year,
   18·1 - 19% with yeast count 10/gm - OK for 1 year.

• Other points:

   • to be safe from fermentation heat honey to 140°F for 30 minutes in order to kill the yeasts but it does partially destroy some of the aromas in the honey.
   • rape honey needs particular care. The beekeeper is tempted to remove the crop as soon as possible to prevent granulation in the comb and it is seldom heat treated because it is always prepared as granulated honey.
   • Badly fermented honey is most unsuitable for feeding to the bees and unsuitable for sale.

3.7 Use of nectar, honey and water by the honeybee colony.

• There are two inputs to the hive, nectar and water (ignoring propolis and pollen):

   • water - used for cooling and humidity control of the brood nest and the dilution of honey.
   • nectar - used for converting to honey for food and storing.

• Food transfer from bee to bee is a continuous process within the colony and the average honey sac content is 50% sugar and 50% water. This average content is important. It is the right mixture for digestion by the bee. The use of nectar, honey and water is shown in the Honey Usage' diagram [appendix 4].

• Points of interest on the diagram:

  • balance of 50/50 is always maintained in bee's 'stomach'.
  • nectar flowing in: no water required; manipulation ('ripening') loses 15% approx. of
                          water. Excess above colony requirements is stored after full ripening
                          (evaporation).
  • no nectar flowing in: honey reserves must be used and water is required for dilution of the
                          honey from 80/20 to 50/50.Foragers will be water carriers.

• During wintering water is required to dilute stored honey. Where does this come from?
Winter water flights are very limited (cold weather). Other sources of water are:

  - CO_2 + water vapour from trachea. H_2O from this water vapour
  condenses within the hive,
  - honey uncapped and being hygroscopic it absorbs some water on the
  surface,
  - re-absorption through the small intestine.

3.8 How pollen is collected, carried back to the hive and stored.

• The colony needs a fertile queen and pheromone from open brood to stimulate the foraging bees to
collect pollen. Returning foragers recruit further foragers by dancing and thereby indicate the
position of the source; the type of pollen is known by the aroma of the pollen on the bees legs.

• The number of pollen foragers can vary between wide limits depending on the colony requirements
(eg. a few percent to as much as 90%).

• The average pollen load (both pellets) = 12 - 30mgs.
The average trips per day = 6 - 8.
Total collected in one year ≈ 100lbs per colony.
Amount required to raise one adult bee = 70 - 150mgs.

• When foraging, the bee alights on a flower and moving quickly bites the anthers of the stamen with
her mandibles in order to dislodge the pollen grains. These pollen grains attach themselves to the
plumose hairs which cover the whole of the exoskeleton. Then she leaves the flower and hovers
nearby to clean the pollen from her body and to load it into her 'pollen baskets'.

  • The front legs: by means of stiff hairs collect pollen, moistened by honey or nectar
    deposited on the front legs, from the head and first thoracic segment.
  • The middle legs: collect the pollen from the first legs and the rest of the thorax particularly
    the ventral side which is then passed on to the inner side of the barsitarsi of the hind legs.
  • The hind legs: clean the abdomen and when sufficient pollen is collected on the inner
    surface of the barsitarsi, these surfaces are raked by the 'pollen rake' at the bottom of the
    tibia of the other hind leg. The pollen is forced as a paste onto the flat surface of the auricle
    which is bevelled upwards and outwards. The tarsus closes against the tibia and the pollen
    is squeezed upwards and outwards onto the outside surface of the tibia. It is held in place
    here by the hairs on the corbicula (note the single hair acting as a pin through the load).
    One full load of two pellets represents approximately 100 flowers visited for a plentiful
    supply eg. dandelion when it is yielding well.

• Storage of pollen. When the pollen forager returns to the hive with a load, she selects a cell near to the unsealed brood, grasps the edge of the cell with her forelegs, arches her abdomen so that the posterior end rests on the opposite side of the cell. The hind legs hang into the cell and the middle legs are used to push the pollen loads off the rear legs into the cell. The forager departs more or less immediately for another load. A house bee now comes along and breaks up the pollen and presses it firmly into the bottom of the cell with her mandibles. Honey or nectar is added to the pollen mass; it becomes darker, has a higher sugar content and is known as 'bees bread'.

• The packed pollen can be fed to the brood or house bees (for producing brood food) or the cell can be filled with further loads topped off with honey and sealed with a wax capping for winter stores.

• All pollen storage is adjacent to and around the brood nest where it is required for use.

• It should be noted that after pollen has been collected by the bee, it is no longer viable for plant reproduction.

3.9 The importance of pollen in the nutrition of the honeybee.

• Pollen is the male germ cell of flowering plants (angiosperms); it has two major uses:

  • it is the principal source of protein, fat and minerals in the honeybee diet,
  • it can provide a surplus product from the apiary.

• Pollen demand in the colony is related to the amount of unsealed brood. Bees cannot rear brood without pollen because the nurse bees would not be able to produce brood food from the hypopharyngeal glands. A strong colony will collect c. 50 - 100lbs. during a season.

  • It requires 70 - 150mg. of pollen to rear one adult bee.
  • About 200,000 bees are reared during a season thus accounting for more than 50% of the income.
  • The balance is used by the adult bees preparing for winter (increasing their fat bodies) and/or stored in the comb for use early the following year before new supplies become available.
  • Note the weight of a worker bee ≈ 90mg. and that 1lb. of bees contains c. 5000 bees.

• Pollen is rich in protein and is essential for body building material for growth/development and for the repair of worn out tissue. It also has the very important function of stimulating the development of the hypopharyngeal glands and the fat bodies of the winter bee. The protein content varies between different pollen types and also from flower to flower in the same foraging area. A protein content of c. 35% is typical of a high protein pollen eg. beans. Bees can discriminate between pollens by colour and odour; they cannot distinguish between the quality (protein content) of various pollens.

  • Pollen contains: - proteins   7 - 35%
                      - lipids (fats/oils) 1 - 14%
                      - amino acids
                      - carbohydrates
                      - minerals   1 - 5%
                      - vitamins

```
                    - enzymes
                    - water          7 - 15%
                    - sugars         25 - 48%
```

There are wide variations in the content of different pollens and the bee more than likely receives a balance diet due the variety of pollens collected and used.

• The use of pollen for brood rearing:

   • Worker larvae are fed brood food only from 0 to 3d. and then on 4th and 5th day with pollen, honey and brood food.
   • Queens both adult and larvae are fed exclusively on royal jelly.
   • After emergence of the worker bee, pollen is essential for it to reach maturity in a healthy state. It depends on pollen for its orderly development of its glandular system while it is a house bee.

• In areas where natural pollen is in short supply, particularly in the spring, pollen patties can be fed to colonies to stimulate spring build up. Pollen shortage often occurs where colonies are foraging on honeydew in pine woods.

3.10 The collection and use of propolis by the honeybee.

• Propolis (pro = before, polis = city) is a resinous gum found on trees and other plants. It was given the name because of the instinct of many strains of bee to build curtains of propolis to restrict the entrance to their nests.

• It is collected with great difficulty by foraging bees in warm weather by:

   • biting with the mandibles,
   • kneading a small piece bitten off with the mandibles,
   • transferring the piece with the 2nd leg to the pollen basket direct,
   • patting it into position again with the 2nd leg,
   • repeating the process to the other pollen basket on the other leg.

• Load is approx. ? mg. and takes 30 - 60 mins. to collect. When the forager returns to the hive, it needs assistance to unload and the unloading takes place at or very near the site it is to be used:

   • unloaded by another worker bee,
   • it bites and pulls the propolis off the carrier bee and puts it in place,
   • the 'cementing bee' may mix wax with the propolis,
   • forager pats the remaining part of the load smooth again,
   • forager is freed of its load in about 1 hour (or several depending on its use in the hive).

• There are only a few propolis foragers in each colony (about 0.5% has been quoted) but this must vary considerably with the strain of bee some of which collect enormous quantities (eg. Caucasians).

• The uses of propolis by the honeybee are:

   • to fill cracks in the hive,

- to reduce openings (eg. the entrance),
- to smooth the interior of the hive,
- to varnish the interior of brood cells,
- to strengthen comb attachments,
- to cover intruders, when they are dead, too large to carry out of the hive (eg. a mouse).

• The uses of propolis by man are:

- medicinal and vetinary work,
- as a varnish or shellac (eg. on violins).

• The beekeeper dislikes propolis because:
- it sticks to the hands, clothing, etc.,
- contaminates beeswax,
- difficult to remove from sections for sale,
- makes frames and other moveable parts of the hive difficult to remove. It is however useful for migratory work by keeping the hive parts stuck together.

** ** ** **

4.0 FLORA and POLLINATION

4.1 The main nectar and pollen producing plants of the British Isles and their flowering periods.

COMMON NAME	PROPER NAME	FAMILY	NECTAR or POLLEN
February/March			
Snowdrop	Galanthus nivalis	Amaryllidaceae	P
Crocus	Crocus spp.	Iridaceae	P
Gorse	Ulex europaeus	Leguminosae	P
Hazel	Corylus avellana	Corylaceae	P
Willow(goat)	Salix caprea	Salicaceae	P
Yew	Taxus baccata	Taxaceae	P
March/April/May			
Blackthorn(sloe)	Prunus spinosa	Rosaceae	N+P
Dandelion	Taraxacum spp *	Asteraceae	N+P
Gooseberry	Ribes uva-crispa	Grossulariaceae	N
Currants	Ribes spp	Grossulariaceae	N
Rape	Brassica napus	Cruciferae	N+P
Top fruit	**	Rosaceae	N+P
Bluebell	Endymion non-scriptus	Liliaceae	N+P
Sycamore	Acer pseudoplatanus	Aceraceae	N+P
H. Chestnut (Wh.)	Aesculus hippocastanum	Hippocastanaceae	N+P
(Red)	Aesculus carnea	Hippocastanaceae	N+P
Hawthorn	Crataegus monogyna	Rosaceae	N+P
Holly	Ilex aquifolium	Aquifoliaceae	N+P
Mountain ash	Sorbus aucuparia	Rosaceae	N+P
Laurel	Prunus laurocerasus	Rosaceae	N+P
June/July/August			
Poppy	Papaver rhoeas	Papaveraceae	P
Thistle	Cirsium arvense	Asteraceae	N+P
Hogweed	Heracleum sphondylium	Umbelliferae	N+P
Field bean	Vicia faba	Leguminosae	N+P
Raspberry	Rubus idaeus	Rosaceae	N+P
White clover	Trifolium repens	Leguminosae	N+P
Charlock	Sinapis arvensis	Cruciferae	N+P
Runner bean	Phaseolus multiflorus	Leguminosae	N+P
Lime	Tilia vulgaris	Tiliaceae	N+P
Blackberry	Rubus fructicosus	Rosaceae	N+P
Willow herb	Epilobium augustifolium	Onagraceae	N+P
August/September			
Evening primrose	Oenothera biennis	Onagraceae	P
Bell heather	Erica cinerea	Ericaceae	N+P
Ling	Calluna vulgaris	Ericaceae	N+P
Old man's beard	Clematis vitalba	Ranunculaceae	N+P
September/October			
Ivy	Hedera helix	Araliaceae	N+P
Michaelmas daisy	Aster novi-belgii	Asteraceae	P

* - Taraxacum officinale is the common dandelion; spp. denotes many species, in this case many species of dandelion.

** - Top fruit include apple, pear, cherry, plum, etc.; all are Rosaceae.

• For the Intermediate examination it has not been formally stated, by the BBKA in the syllabus, that it is necessary to know the botanical names and family of the plants concerned. We believe that it is most desirable that at this level a familiarity with the correct terminology should be developed and therefore these have been included (eg.Ragwort, Senecio jacobaea, family Compositae. The family is now re-named Asteraceae. In Scotland it has a common name: Stinkywilly.).

• It is extremely difficult to define what are the main plants throughout the country because there are many local variations and the student or candidate should be familiar with his local flora; eg. winter aconite (Eranthis hyemalis of the buttercup family Ranunculaceae) is a prime source of pollen in the spring in some parts of Devon and for the last two years our own bees have worked Lesser Celandine (Ranunculus ficaria) at this time of the year. When the authors lived in Sussex, neither of these two plants were worked by the bees in the spring because they were not available.

• In section 1.12 a list of some of the bee forage plants were listed; in this section some of these have been omitted and others included in order to provide a broader spectrum.

• It is to be noted that the labiates (eg. mint, thyme, rosemary, lavender, etc.) are all very attractive to bees but in UK they do not rate as major sources of forage (eg. thyme produces a crop of honey in Greece and Malta).

4.2 The process of pollination and fertilisation of a flowering plant.

• Pollination: is defined as the transfer of pollen from the anthers of a flower to the stigma of that flower or another flower on the same plant/tree or another plant/tree.

• Fertilisation: is defined as the union of the male and female gametes which occurs after pollination.

• The essential organs of a sexually reproducing flower (angiosperm) are as follows [see diagram of flower in appendix 4]:

  • stamen - has a stalk like filament terminated at the distal end by the anther which produces the pollen grains (male gametes). There can be many stamens in a floret in either one or two whorls.
  • pistil - is a long tube called the style terminated at its distal end by the stigma. At the inner end it is terminated in the basal ovary which contains the ovules (female gametes). The stigma has the function of capturing the pollen grains and providing a suitable surface for them to germinate, when it is in this state it is said to be receptive. In any one floret there is only one style and stigma.

The stigma and the stamens are surrounded by petals and then by sepals, the whole being carried on the stem of the plant. The number of stamens, petals and sepals are a useful guide to the identification of the plant.

- The pollen grain contains:
    - a tube nucleus which is responsible for the growth of the tube from the stigma down through the style into the ovary and into the ovules.
    - one or two sperm nuclei responsible for fertilising the ovaries and which follow down the tube behind the tube nucleus as the tube is being formed.

- The process of pollination is extremely simple; it is the transfer of the pollen grain from the anther to the stigma when the stigma is receptive and when the pollen is viable (alive).

- When the flower (taking an apple as an example) opens it generally remains open for c. 5 or 6 days before the petals fall. About day 7 or 8 the flower aborts and if fertilisation has not been completed before this time then no fruit will set. The critical time is up to petal fall. The stigma is receptive between the time the flower opens and about day 3 and pollination must occur during this time. After germination on the stigma it takes about 5 days for the pollen tube to grow and fertilisation to take place. It is for this reason that the deadline for pollination is day 3 because adding 5 days for pollen tube growth brings the time to day 8 the flower aborting time. In warmer climates [eg. Mediterranean] the pollen tube grows faster, typically 2 or 3 days, making the fertilisation process more reliable.

- After fertilisation the ovules become the seeds of the fruit and the ovary becomes the actual fruit as we know it.

- Successful pollination and fertilisation is dependent on a number of factors:

    - the availability of an adequate number of pollinators (eg. bees),
    - the weather conditions to allow the pollinators to fly,
    - the temperature to produce nectar to attract the pollinators,
    - humidity to be relatively low so that the pollen is viable; often large quantities of dead pollen are transferred by the bees to no avail,
    - the temperature must be high enough to complete the pollen tube growth before the flower aborts.

- The following terms should be noted:

    - self pollination - same flowers of identical genetic material,
    - cross pollination - transfer of pollen which is not identical genetic material,
    - self fruitful - capable of fertilisation with its own pollen,
    - self unfruitful - do not become fertilised when self pollinated,
    - cross fruitful - become fertilised when cross pollinated,
    - cross unfruitful - non compatible when cross pollinated.

- Bees pollinate, they do not fertilise. When the flowers open and the stigma is receptive, nectar is usually secreted as an attractant to the pollinators. The aroma of the flower is also attractive to the bee at a range of a few feet. There are a wide range of estimates on the number of pollen grains that can be collected on the plumose hairs of the honeybee; these range from 50,000 to 5,000,000 depending on which source is quoted.

- There are a variety of methods of pollination which have been evolved by the flowers; insects, wind, animals, birds, water for example. The two following terms should be noted:
    - anemophilous - wind pollinated
    - entomophilous - pollinated by insects.

4.3 An account of the honeybee as a pollinating insect and its usefulness to farmers and growers.

• Honeybees evolved from wasp like ancestors c. 20 million years ago, at the time the flowering plants were developing. A mutually beneficial relationship has developed between sexually reproducing flowers and the honeybees. The bees provide for cross pollination of plants thereby ensuring a greater variability in the offspring than self pollination. The plants in turn provide the bees with a reward of nectar. Other less important pollinators include flies, beetles, butterflies, bats and wind.

• Bees have evolved branched hairs which can carry large numbers of pollen grains, intricate pollen baskets, specialised mouthparts, honey sac for transporting nectar, beeswax for building comb to store honey and pollen, specialised behaviour for communication, etc. all of which are related to their association with angiosperm plants.

• Flowers attract bees by:

  • colour (flowers reflect UV light)
  • scent
  • nectary guides
  • shape

Flowers themselves prevent self fertilisation (self unfruitful) eg. Most apple varieties require pollen of another variety for successful fertilisation. Pear pollen is too sticky for it to be wind pollinated.

• In UK most of the crops requiring pollination flower in the spring, the crop maturing in the summer and being ready for harvesting late summer/early autumn. The honeybee is the only pollinator which over-winters as a colony and which is available in the large numbers required early in the year. Practically all the solitary wasps and bees together with the bumble bees and social wasps start the season with only a fertilised queen who starts reproducing in the spring. This is the major reason that the honeybee is so valuable in the UK climate for pollination purposes.

• Because the honeybee is kept in hives and managed by man, whole colonies in very large numbers can be transported and sited in the crops to be pollinated. The colonies can be distributed throughout the crop to the best advantage and provided in the optimum numbers required. The colony density for fruit pollination is about 2 - 6 colonies per acre.

• Honeybees are polytropic but constant to one plant while they are foraging (an advantage on most crops for pollination purposes). A knowledge of bee behaviour is important for honey production and also for pollination:

  • a colony with 2·5 miles flight radius has access to 12500 acres
  • most crops can support 1 colony/acre for nectar secretion (eg. rape)
  • bee activities are effected by the weather:
      - T 100°F [38°C] and no bees will be flying,
      - they fly and forage in winds up to 15 mph,
      - above 15 mph activity decreases and stops at 21-25 mph,
      - cool cloudy weather greatly reduces bee flights,
      - T > 13°C [55°F] are required for pollination (flower requirement),
      - colonies for pollination must be strong and have large brood nests to stimulate

pollen foraging [a minimum of 5 or 6 frames of brood is usually acceptable but the pollination contract should state a definite figure].
- feeding sugar syrup may be necessary to stimulate brood rearing.

• It has been shown that crops such as rape (wind pollinated) provide a better set / greater yield if colonies of bees are available in correct numbers and correctly distributed while the crop is in flower. Any crop which relies on insect pollination produces better yields when bees are provided as pollinators. The UN organisation, FAO, has estimated that about one third human diet of the western world comes directly or indirectly from insect pollinated foods. About 90 crops in USA depend on bees for pollination.

** ** ** **

5.0 DISEASE and POISONING

A few introductory words related to disease which do not appear in any of the set reading for the intermediate examination. The words signs and symptoms are very often confused; even in the BBKA examination papers if you look carefully at some of the past ones! Signs of a disease are something you can see, whereas symptoms are something the patient feels or suffers. Pathogens (the agents causing disease) related to bee diseases vary widely in size and in their form, a brief list is shown below:

- Virus: the smallest pathogen (6 - 400nm) and, at present, there is no cure for any of the viral infections [eg.Sacbrood, Chronic Bee Paralysis Virus (CBPV), Black Queen Cell Virus]. A virus attacks the cell and any treatment to kill the virus would kill the cell itself. An electron microscope is required to observe them.
- Bacteria: the next largest (0·1 - c.20μm), most can be seen under the light microscope and infections can be treated with antibiotics [eg. EFB and AFB].
- Fungi: fungal spores are about the size of bacteria [eg.Chalk brood and Stone brood]. The spores can be killed [eg. with acetic acid fumes] but the disease, once established in the larvae cannot be treated.
- Protozoa: again about the size of bacteria (5 - 12μm) and readily seen under a light microscope [eg. Nosema and Malpighamoeba] with a magnification of about 400.
- Mites: all with 8 legs of the spider family (c.120μm - 2mm) readily seen with a hand lens [eg. Acarine (Acarapis woodi) infested trachea and Varroa jacobsoni].
- Flies: (1 - 1·5mm) readily seen by eye [eg. Braula coeca, not technically a pathogen] and always have 6 legs; the bee louse (Braula coeca) is a wingless fly.
- Moths: the wax moths are about the largest pathogens (?) associated with bees.

Now to get the measurements into perspective:

$$1 \text{ metre} \div 1000 = 1\text{mm (milli-metre)}$$
$$1 \text{ mm} \div 1000 = 1 \text{ μm (micro-metre or micron)}$$
$$1 \text{ μm} \div 1000 = 1 \text{ nm (nano-metre)}$$

5.1 Field diagnosis of AFB and EFB and the signs of the two diseases.

• ADAS leaflet # P306 'Foul brood of bees: recognition and control' should be obtained [free of charge].

• Both diseases are diseases of the brood and there are no signs associated with the adult bees in an infected colony. In order to diagnose either in the field, it is necessary to open up the colony and examine the combs containing brood. To do this properly it is necessary to shake the bees off the comb before examining it, leaving no more than a few bees on the comb. The reason for this is that in the early stages only an odd cell or two will be exhibiting the tell-tale signs. This important aspect of searching for the diseases is frequently overlooked and inadequately expressed in much of the literature. There is a right and a wrong way of shaking bees off combs, the objective is to rid the comb of bees and keep them in the hive (not flying around the apiary); therefore raise the comb slightly and shake it sharply in the brood chamber without jarring the rest of the colony.

• In order to diagnose the diseases in the field it is easier to remember the signs if one has an understanding of the progress of the diseases:

- AFB [American Foul Brood]: The larva is fed the AFB spores with the larval food. The spores germinate in the ventriculus and the larva dies **after** the cell is sealed. The germinated spores break through the wall of the ventriculus into the haemolymph and the larva dies of **septicaemia** * ; then the whole larval form disintegrates, melts down, becomes thick and sticky and finally dries to a hard scale on the lower angle of the cell. During this deathly saga the colour changes from white to black. It is most important to note that prior to the sealing of the cell, the larvae appear to be perfectly healthy. * Septicaemia - is the circulation and multiplication of micro-organisms in the blood.

- EFB [European Foul Brood]: The larva is again fed the pathogen, this time a bacteria which is not spore forming as was AFB, which multiplies in the ventriculus by using the larval food and the larva dies before the cell is sealed due to starvation. It dies at about day 3 or 4, so it is quite large when it dies. A dead larva is not sealed by the bees and is removed. During the starvation period the larva contorts into unnatural shapes in its cell and changes colour from a pearly white to cream to yellow to light browny green (colours are difficult to describe in words; any deviation from the pearly shiny white must be regarded with suspicion). When the bees remove the dead larvae, they are removed in one piece and they are either there to see or else the signs have been removed.

- Signs of AFB (caused by the spore forming bacteria 'Bacillus larvae'):

  - open brood - no signs,
  - sealed brood, many signs as follows:
    - After the larva dies, the domed cells become moist and darken in colour.
    - Cappings then sink and become concave (still moist and discoloured).
    - Holes appear in the cappings (ie. perforated).
    - Matchstick pushed through sunken capping to test for roping of the contents. Length of 'rope' between 1 and 2 cms. This roping is considered to be a positive identification of the disease. Colour of cell deposit is from light brown to nearly black. The roping test can only be done between the time the larva has 'melted' and the melt thickened slightly and before it has dried too much to be sticky.
    - The remains dry out on the lower angle of the cell and form a hard black scale. By the time the scale is formed, the bees have uncapped the cell completely and tried to remove the scale. In order to see the scale the comb must be held at an angle with the top bar closest to you and the bottom of the frame away from the body (angle about 45°to the vertical). Good light is essential, some books say from the back while others say from the front; we think either is acceptable depending on whether you are in or outdoors. In the early stages of the infection, possibly only one or two cells may have scale and this is why it is so important to clear the frames of all the bees when doing an inspection for foul brood.
    - Brood combs which have a 'pepperpot' appearance (ie. empty cells among sealed brood) should be treated with suspicion and examined closely for any sign of scale.
  - AFB infections have no smell; many books indicate a foul odour. Bacillus larvae when sporulating releases an antibiotic preventing any secondary infections. If an offensive smell is present it will be due to secondary infections of some other cause or the confusion may arise because when the bacteria are in the rod form all the cells are sealed and no odour can be released. Rely on visual signs not odour for AFB.
  - AFB is easily identified visually in the field, however, it can be confirmed if necessary by laboratory tests usually on a piece of scale from the comb.

• Signs of EFB (non-spore forming bacteria 'Melissococcus pluton'):

  • Sealed brood - no signs (the larva dies before sealing).
  • Open brood - the signs are as follows:
    - Larvae are usually in unnatural contorted positions in the cells; twisted spirally or flattened out lengthwise (nb. stomach ache is a good analogy).
    - The colour changes from a pearly white of a healthy larva to dull cream, to light brown and eventually a greeny hue. The colour change should be associated with the unnatural positions.
    - The dead larvae have a melted down appearance but still have a larvae like shape.
    - Again EFB does not itself smell. However very often an offensive smell is present on combs with EFB infected larvae; these are secondary infections often associated with EFB and are another indication that the disease may be present (2 common secondary infections are from Bacillus alvei and Bacterium eurydice).
  • EFB is very difficult to diagnose positively in the field for the following reasons:
    - The larvae are removed quickly from the hive once they are dead so the evidence is often removed and not there for the beekeeper to see.
    - Any diseased larvae can be confused with other brood diseases, such as Sacbrood or Neglected drone brood, unless the beekeeper is very experienced. Most Foul Brood Officers will remove a frame with dead and dying larvae for laboratory analysis.
    - The best time to look for EFB is when the brood outnumbers the adult bees (see colony population cycle) in the spring about mid April to early May. At this time the chances of spotting the diseased larvae are greater because the house bees are fully stretched under these conditions.

5.2 Spread of foul brood infections from colony to colony.

• Both diseases are originally transmitted to the larvae through feeding the spores or bacteria in the larval food by the nurse bees.

• In the case of AFB the larva dies after the cell is sealed. The cell is infected with spores due to larval defaecation and later by the melted down remains and eventually the hard scale. Any infected cell has millions of latent spores. House bees try to clean the cells and their mouth parts become infected, the spores being passed on to the nurse bees during food transfer. The nurse bees then infect the young larvae.

• The mechanism for EFB is similar but in this case the larva dies before the cell is sealed and dies as a result of starvation. If the larva is removed, the infection is removed with it; it is the infected larvae which do not die that spread the disease within the colony. Again the cell is infected by larval defaecation, and passed on to the house bees during cell cleaning and food transfer. Because every cell is not infected it is not as contagious as AFB and this is one reason why EFB can appear and disappear in a colony from time to time. It can be very elusive.

• The spread within the colony is beyond the control of the beekeeper, but the spread between colonies is very much in his hands. It is spread as follows:

  • By robbing; likely when the infected colony becomes weak and is then robbed by strong healthy colonies either within an apiary or between apiaries.

- By drifting; adjacent hives in an apiary which are not orientated in different directions.
- By feeding infected honey; bees should always be fed sugar syrup.
- By bees gaining access to infected honey, combs, wax and propolis left around the apiary or within foraging distance.
- From appliances (eg. hive tools, extractors, etc.).
- From second hand infected equipment (combs, hive parts, etc.).
- From swarms of unknown origin.
- By the exchange of combs between hives.
- By the purchase of bees from a doubtful source.

5.3 Action to be taken when AFB or EFB is found, including treatments and sterilisation of equipment.

- Two avenues exist:

  - when found by the beekeeper - report findings to MAFF immediately and follow their instructions,
  - when suspected by the beekeeper and found by MAFF (FBO) - follow their instructions implicitly.

- If AFB is diagnosed then the treatment is always to destroy the colony; the beekeeper being served a notice which he must sign. If EFB is diagnosed then the condition may be treated with antibiotics but only at the discretion of the FBO. Individual beekeepers are not allowed to treat their own colonies and it is against the law to do so. The work of destroying a colony falls to the owner but must be supervised by the FBO. The FBO usually administers any antibiotic treatment for the owner which is followed up at a later date by further inspection of the colony and, more than likely, any other colonies in the apiary and or adjacent apiaries. In both diseases a standstill order is put into effect banning the movement of colonies and equipment into or out of the apiary concerned. The standstill order is operative until such time as it is cancelled by the FBO in writing.

- Colony destruction treatment for AFB:

  - Colony may only be destroyed after dark when all bees of that colony have returned and stopped flying.
  - Before the evening seal all openings (except the entrance which is reduced to c. 2") and put zinc gauze over the feed hole.
  - When the bees have stopped flying, block the entrance securely (eg. clod of earth) and then pour ½ pint petrol into the colony through the feed hole. The bees will be dead in a few minutes. Leave for 10 mins.
  - Dig hole 3 ft. square by 3 ft. deep. This can be done earlier and in a position not too far away to minimise carrying equipment to the burning site at the bottom of the hole.
  - Prepare starter paper + 2 or 3 combs and frames pyramid fashion.
  - Set alight and burn all hive contents including combs, frames and quilts if these are used.
  - Scrape all boxes, floorboard and crown board free of wax and propolis into the fire.
  - When completely burnt out back fill immediately.
  - Scorch all hive parts with a blow lamp to a coffee brown colour paying particular attention to corners and cracks in the woodwork and to the queen excluder.
  - Disinfect any other appliances eg. smoker, feeder, hive tool, etc. in solution of:

- 1lb washing soda,
- ½lb bleaching powder,
- 1gallon warm water.

Use while warm and rinse in clean water before drying (note that the solution is caustic - take care).
- Finally obtain a destruction certificate from the FBO to substantiate your insurance claim on BDI.

5.4 Method of detecting Varroa jacobsoni and its differentiation from Braula coeca.

- The physical differences between Braula coeca and Varroa jacobsoni are as follows:

  - **Braula coeca**: ellipse shaped c.1 - 2 mm with 6 legs, coloured reddish brown. It is a wingless fly. Initially it is white and takes about 12 hours to turn colour. The head and posterior end of its abdomen are on the ends of the major axis of the ellipse, the legs are on the sides associated with the ends of the minor axis of the ellipse looking down on the dorsal side. Easily seen by eye riding on worker bees and very often the queen is infested. Causes no harm to queen or bees but the larvae spoil capped honey comb with fine tunnels in the cappings.
  - **Varroa jacobsoni**: also ellipse shaped c. 1·1 × 1·7mm. with 8 legs, coloured reddish brown the same as the Braula coeca. The legs are on the ventral side and cannot be seen when it is viewed looking down on the dorsal side. It travels 'blunt end first' its legs being on the sides associated with the ends of the major axis. It is an arachnida and is in the spider class in the animal kingdom not the insect class. These mites are difficult to detect as they feed on the haemolymph by piercing the membrane between the abdominal segments on the adult bee and breed in the capped brood cells.

- The recommended method of detection at the present time is the tobacco smoke method fully described in MAFF pamphlet 936 published 1985, which is available free of charge. The leaflet advises testing in October each year and sending the debris to Luddington for analysis. There is no charge for this 'Varroa service'.

5.5 The major legal requirements relating to Foul Brood, Varroasis and the importation of bees applicable to England and Wales.

- There are 3 major pieces of legislation as follows:

  - The Bees Act 1980,
  - The Bee Diseases Control Order 1982, S.1.107 (AFB, EFB, Varroasis).
  - The Importation of Bees Order 1980.

- AFB, EFB, Varroasis: the major points in the legislation are as follows:

  - Notification of disease: the beekeeper who suspects disease **shall** notify with all speed and **shall not** move bees, hive, etc. until the authorised person (FBO) has examined them.
  - Precautions against spread of infection:

- Authorised person **may** take samples of combs (for AFB, EFB) and combs plus bees and debris (for Varroasis).
- Authorised person, if he suspects disease, **shall** serve a notice prohibiting removal except by licence.
- Authorised person may mark any hive or appliance for identification purposes.
- If samples are positive, the authorised person **shall** serve a notice.

- AFB:

  - If the authorised person and the beekeeper agree the disease, beekeeper signifies his agreement by signing notice.
  - Notice requires destruction, details of which are specified.
  - Destruction to be supervised.
  - Notice in force until date subsequently notified.

- EFB:

  - Same as AFB except colony **may** be treated in specified manner in lieu of destruction but again under supervision.
  - The authorised person may serve notice on any other beekeeper whose bees may likely to come into contact with diseased bees.

- Varroasis:

  - Same as EFB but also require isolation period as stated in the notice.
  - Declare an area as infected by being published publicly (in the local press?).

- Beekeepers are required to:

  - Notify if disease is suspected.
  - Must not interfere with identity marks on equipment.
  - Provide facilities and information to authorised person.
  - Must not treat bees with drugs that disguise presence of disease.

• It is advisable to obtain a copy of the legislation which is obtainable from HMSO. Note that 'shall' in legal terms is mandatory, whereas 'may' is not mandatory.

• The Importation of Bees Order 1980 prohibits the importation of bees but makes provision for the importation of queens and attendants from Varroa free countries under licence. In practice this is now New Zealand only; and some authorities argue that because of the lack of knowledge about Kashmir Bee Virus, this country should also be banned. Buckfast queens will be available from Teneriffe in 1991; a new scheme set up by Bro. Adam. Note that the order does not prohibit the movement of genetic material such as eggs or sperm.

5.6 An account of the signs of Chalk Brood, Sac Brood, Bald Brood, Varroasis, Addled Brood and Stone Brood.

• **Chalk Brood** (a fungus - 'Ascosphaera apis')

  • Signs:
  - Larvae die after the cell is capped.
  - Occasional cell infected or large areas of brood.
  - Capping removed by bees.
  - Larvae become chalky white, fluffy and swell to fill the hexagonal cells.
  - Larvae then shrink and harden.
  - When infected with two strains of fungus, the colour becomes dark grey or black.
  - Larvae removed by house bees.

- Dead larvae ['mummies'] are found outside the stock or in a badly infected stock on the floor board.
- Common infection in U.K. especially in spring or in newly made nuclei.
- Dry discoloured pollen pellets are sometimes confused with 'mummies' but on closer inspection the layers of pollen of different colours are easily discernible.
- Causes:
  - Presence of spores on the nurse bees, combs and hive parts of an infected hive.
  - Stress creating a condition for the spores to germinate:
    * lower than normal temperature in brood nest,
    * protein shortage,
    * level of CO_2 above 0·5% in the brood nest.
  - Unbalanced colony causing stress as above is due mainly to insufficient bees eg. a newly made nucleus.
- Treatment: nil, (spores killed by acetic acid sterilisation of combs and hive parts). In severe cases, re-queen with a more resistant strain of bee.

Sacbrood (a virus - 'Sacbrood virus')

- Signs:
  - The infection interferes with the moulting process and the final larval moult does not occur (ie. 5th moult after the cell is sealed) with a result the larva does not pupate and dies stretched out in its cell.
  - Cells containing dead larvae are uncapped by the bees.
  - As the moulting fluid collects between the body and the unshed skin, the colour changes from pearly white to pale yellow.
  - After death, a few days later, the colour changes to dark brown; the head changing colour first.
  - Finally, the larva dries down to a flattened shape with a slightly upturned head (Chinese slipper effect).
  - In the yellowy colour stage it can be confused with EFB and in the Chinese slipper stage it can be confused with AFB.
  - Disease is very common in UK (30% of colonies likely to have it).
  - Likely to be noticed more often in the spring and early summer when ratio of brood to bees is high.
- Treatment: nil, re-queening is said to be effective in severe cases.

Bald Brood (Greater wax moth - 'Galleria mellonella')

- Signs:
  - A Galleria mellonella larva hatches among the brood and chews its way through brood cappings in a straight line. The bees remove the silk tunnels and leave the bee larvae bare which are not re capped.
  - Often the capping is not quite cleared at the angles of the hexagonal giving a slightly raised appearance at the edges.
  - If due to wax moth, the bare cells are always in a straight line where the wax moth larva has eaten its way forwards.
- Causes:
  - As above, wax moth damage.
  - Genetic trait of some strains of bee. Often small patches of brood are uncapped.
- Treatment: nil, the brood emerges normally in the case of the genetic fault but is sometimes crippled with deformed wings and legs due to faecal pellets from the wax moth larva.

- **Addled Brood**

    - Most of what used to be called Addled Brood is now identified as Sacbrood. Addled Brood now generally seems to be applied to any brood disease that cannot be positively identified.
    - At Luddington, the diagnostic services (examining 1000's of combs) find many which do not match the known diseases.
    - The many viral diseases and conditions due to hereditary faults prompts the question "Is there such a thing as Addled Brood?"
    - Signs: not defined.
    - Cause: generally unknown.

- **Stone Brood** (a fungus, either 'Aspergillus flavus' or 'A. fumigatus')

    - Signs:
        - The disease is extremely rare in UK.
        - Most larvae generally die after they have been capped prior to pupation.
        - Larvae with Stone Brood may be either capped or uncapped.
        - A. flavus (the most common but both are rare) has yellow - green appearance.
        - A. fumigatus has grey - green appearance.
        - The dead larvae are similar in appearance to Chalk Brood except for the colour.
    - Cause: both fungi are common and occur in the soil and in cereal products (eg. mouldy hay). Probably the spores do not normally germinate in the bee or its environment, thus making it rare in UK.
    - Treatment: some books recommend destruction of the colony.
    - It should be noted that both fungi can effect animals and humans. It is therefore important not to sniff or smell comb with stone brood as respiratory infections are likely to result which are difficult to cure as they are resistant to antibiotic treatments.
    - Aspergillus fumigatus is used to prepare the antibiotic Fumidil 'B' for the treatment of Nosema.

- **Varroasis** ('Varroa jacobsoni', mite of the class Arachnida)

    - Varroasis not only effects the sealed brood where it breeds but it also lives on the adult bee and feeds on its haemolymph at the intersegmental membranes. It can therefore be classed as both a brood disease and an adult bee disease.
    - Signs:
        - It is unlikely that any sign of Varroasis will be apparent until the colony has been infected for about 3 years. The first indications are likely to be a general weakening of the colony.
        - It is very unlikely that the mites will be seen on the adult bees as they generally inhabit the ventral side of the abdomen. The most positive sign is by knock down test using tobacco smoke and collecting the knocked down mites on the floorboard.
        - As the V.jacobsoni breeds in the sealed brood cell (drone preferred), they cannot be seen except by opening up the cells and conducting a proper search.
    - Treatment: nil at present in the UK. It is a notifiable disease and the action to be taken, as directed by MAFF, will depend very much on the circumstances involved for the first infestation.
    - It is encumbent on every beekeeper to make a regular search every year for this scourge which will have serious consequences on colony management if it arrives in UK.

· **Neglected Drone Brood:** this is not listed in the syllabus but should be known to every beekeeper. This is not technically a disease, it is more a condition of the colony (it is not initiated by a pathogen).

- It is caused by a drone laying queen or by a colony with no queen and laying workers.
- Signs:
    - The colony is usually small and will have dwindled, the drones produced in worker cells being in evidence (stunted and mal-formed).
    - Typical raised and domed cappings on worker cells are evident.
    - Because the colony dwindles, the bees eventually neglect drone brood in worker cells which then die of starvation before sealing of the cell. They then start to decompose, lose their normal shape and become discoloured (white to yellow to brown).
    - The decomposing larvae become brown watery mass [which does not 'rope'] and eventually dries to a scale which can be removed by the bees.
- It should be noted that neglected drone brood can be confused with EFB during the discoloured larvae stage and with AFB at the scaling stage.

5.7 An account of the signs of and the recommended treatment for adult bee diseases.

· The adult bee diseases are:

- Acarine,
- Nosema,
- Amoeba,
- Dysentery,
- Varroasis,
- Viral diseases (eg.Paralysis and others).

- **Acarine** ('Acarapis woodi' [Rennie] - a mite in the class Arachnida)

- these mites, [c. 150μ × 65μ require a μscope to see them] were discovered by Dr. Rennie at Aberdeen University as a research project funded by the philanthropist Mr. Wood. It is usual for a new biological discovery to be partially named after the scientist who did the research work. In this case both the scientist and the philanthropist are named. This work on Acarine (called at that time 'the Isle of Wight Disease') was commissioned in 1921 after many colonies had been wiped out in UK.
- the EEC terminology for acarine is acariosis and there are proposals [in 1990] that it should become a otifiable disease.
- Signs:
    - Despite the large number of references to the signs of Acarine in beekeeping literature, Dr. Bailey's work has proved that there are no visible external signs of this disease. The disease has no effect on the flying ability of the bee but it does shorten its life slightly (time not quantified).
    - The following signs are those of Chronic Bee Paralysis Virus:
        * crawling with fluttering wings,
        * clinging to plants and stems near the hive,
        * bloated abdomens,
        * crawlers may be in large numbers,

   * dislocated or partially spread wings (K-wings),
   * huddled together on the top bars or on top of the cluster in the hive.

   - Diagnosis of Acarine can only be confirmed by dissection and microscopic examination of the first thoracic trachea. When the disease is present the trachea will be discoloured and not the normal creamy colour of healthy adult bees. The trachea can be infested either on one or both sides. - It should be noted that there is no correlation between the Acarapis woodi and CBPV and Acarapis woodi has not been proved as a vector for the spread of CBPV. This is very curious because when crawling bees, etc. are found in a colony and the bees are examined, in a large percentage of the cases Acarine will be present.
• Treatment:
   - Folbex (Chlorobenzilate) or Folbex VA (Bromopropylate). Both are equally effective. The VA stands for varroa and acarine. Used in warm weather when the supers have been removed, the bees are flying well and there is no risk of contaminating honey for human consumption. See any text book for method of application. Two strips are usually necessary applied c. one week apart. It is to be noted that Folbex is not now produced and Folbex VA has been withdrawn in early 1990, thereby leaving UK with no approved medicant!
   - Other treatments are used and these include Frow Mixture*, fumes of burning sulphur, methylsalicylate, etc. which are not generally used these days by the modern beekeeper. * Frow Mixture = 2 parts nitrobenzene, 1 part safrol and 2 parts petrol.

• **Nosema** ('Nosema apis' [Zander] a spore forming protozoa)

 • The Nosema spore (6 to 8µ) can only be observed using a compound µscope. It was discovered by Prof. Enoch Zander at Erlangen in the early part of the century. The protozoa multiply in the ventriculus (30 to 50 million spores when infection fully developed) and impair the digestion of pollen thereby shortening the life of the bee. It does not effect the honeybee larvae.
• Signs:
   - Infected bees themselves show no outward signs of the disease.
   - Colonies fail to build up normally in the spring.
   - In badly infected colonies in the early part of the year:
    * exhibit signs of dysentery (soiled combs and soiled entrance),
    * dead bees outside hive entrance (after cleansing flights).
   - Diagnosis of nosema can only be confirmed by µscopic examination.
• Treatment:
   - Fumidil'B' inhibits the spores reproducing in the ventriculus. It does not kill the spores.
   - Autumn treatment: Fumidil'B' followed by spring treatment the following year.
   - Spring treatment: Bailey frame change plus Fumidil'B' [see appendix 10].
   - Fumidil'B' administered in syrup 166mg to one gallon per colony.
   - Good beekeeping practices prevents spread of infection in both the hive and the apiary eg.no squashing of bees during manipulations and prevention of robbing, drifting, etc.
   - Sterilisation of infected comb and hive parts with 80% acetic acid (100ml/brood box for one week). Note that acetic acid should be placed on top of the frames because the fumes are heavier than air and sink to the bottom between the frames.

It is also very corrosive and metal ends should be removed from frames and any remaining metal work should be greased before treatment.
- Other points:
    - Due to the high incidence of nosema in UK. it is virtually essential to monitor twice a year by taking samples in spring and autumn and treating as required.
    - It is understood that using Fumidil'B' regularly as a prophylactic for nosema is unlikely, but not impossible, to produce forms of nosema apis resistant to the antibiotic Fumidil'B' (Prof. L.Heath).
    - It is important to ensure that any Fumidil'B' used for this treatment is not time expired and has been stored in the dark.

- **Amoeba** ('Malpighamoeba mellificae' [Prell] a protozoan amoeba-like parasite which ultimately encysts in the malpighian tubules).

    - The cyst which is found in the malpighian tubules is c. 10 - 12μ and can only be detected using a compound microscope. The cysts germinate, develop and multiply in the ventriculus. The amoeba then make their way into the tubules and eventually form cysts which pass through the small intestine and rectum and are voided in the faeces. The infection seems to have no effect on the colony.
    - Signs:
        - There are no external signs.
    - Treatment:
        - No medicants are available for treatment and putting the colony onto clean comb, as for nosema, and then sterilising the comb and hive parts is the only treatment available.

- **Dysentery** (means to soil the combs or hive with excrement).

    - It does not mean the passing of blood, inflamation of the bowel or presence of infection.
    - Dysentery is not a disease but a condition caused by excess water in the intestine which manifests itself mainly in the winter and can be due to:
        - unripe honey and or late feeding,
        - granulated stores,
        - fermenting stores,
        - brown sugar, raw sugar and acid inverted sugars although the reason is not known,
        - extracted honey granulates in the comb if fed back to the bees.
    - Signs:
        - Fouling of combs, hive parts and around the entrance.
        - When rectum weight = ⅓ weight bee ...comb soiling starts.
        - When rectum weight = ½ weight bee....dysentery is certain.
        - In severe cases in bad weather it can kill a colony, but it is more likely that the bees and the colony are so weakened that the colony succumbs due to viral infections.
    - Treatment:
        - warm thick syrup is said to be helpful.

- **Varroasis** ('Varroa jacobsoni').

    - The mite lives on the adult bee and breeds in the brood cells in the brood nest. It can be classed as both a brood disease as well as an adult bee disease. While it is on the bee it feeds on haemolymph by piercing the intersegmental membrane. The effects on the bee are

not clear but it must have an adverse effect and it is known that the mite is a vector for Acute Bee Paralysis Virus (ABPV).
- Signs:
    - Virtually no external signs. The mites are difficult to detect except by knock down tests in the early stages.
    - The colony is considerably weakened over a 2 or 3 year period and finally it dies out if left to its own devices.
- Treatment:
    - Treatment in UK is not allowed. If varroasis is found, it must be reported and MAFF will decide on the treatment or course of action.
    - Much work is being undertaken on treatment for this disease both chemical, biotechnical and biological. There are no preferred methods of treatment at the time of writing. It is likely that it will be necessary to combat the disease with a variety of treatments, particularly the chemical ones, to ensure that no resistant strains are developed by the continual use of one particular medicant.

- **Viral diseases.**

    - Many of the viral infections of the honeybee are associated only with its adult form such as Chronic Bee Paralysis Virus (CBPV), Acute Bee Paralysis Virus (ABPV), Black Queen Cell Virus, Bee Virus Y, Filamentous Virus, etc. They could be classed as adult bee diseases.
    - Conversely there are few viral infections associated with brood diseases such as Sacbrood Virus.

5.8 Chronic Bee Paralysis Virus (CBPV) and Black Queen Cell Virus.

• It should be noted that there are no cures for any viral infection, they are immune from any antibiotic treatment. Viruses only multiply in living cells of their hosts and any medicant which kills the virus would kill the cell and the host. CBPV is responsible for the death of most colonies that succumb to the adult bee diseases. Nosema, acarine, etc. do not kill off the colony, they only weaken it and thereby allow the viral infection to take over. It is for this reason that Dr. Bailey considers that it was not acarine (the I.O.W. disease) that killed off so many colonies at the early part of the century but viral infections; in all probability CBPV which of course had not been identified at that time. An electron microscope is required to see the viruses and to identify them. This type of microscope had not been invented at that time. It is now clear, from work mainly undertaken at Rothamsted, that there is an association between many of the viruses and diseases caused by other pathogens.

- **CBPV.**

    - This virus disease has two distinct syndromes, type 1 and type 2; the signs are quite different.
    - Signs syndrome 1:
        - abnormal trembling of wings and body,
        - bees fail to fly and often crawl on the ground and up plant stems,
        - sometimes the crawling bees are in masses of thousands,
        - huddle together on top of the cluster or on top bars,
        - often have bloated elongated abdomens (due to bloated honey sac),
        - partially spread or dislocated wings.

- Signs syndrome 2:
  - affected bees can fly but they are almost hairless,
  - bees appear dark or black and being hairless appear smaller,
  - relatively broad abdomen,
  - suffer nibbling attacks by older bees in the colony; this may be the cause of hairlessness,
  - they are hindered at the entrance by the guards,
  - a few days after infection trembling sets in then flightlessness and then they soon die.
- Both syndromes can exist in the same colony but it is usual for either one or the other to predominate. Hereditary factors are believed to have some bearing on the susceptibility to the disease.
- It will be clear that most of the signs of syndrome 1 are those stated in much of the literature to be those of acarine but in fact are CBPV.
- In the authors' experience most colonies terminally weakened with nosema or acarine seem to exhibit signs of both syndromes particularly clustering on top bars and continual trembling.

- **Black queen cell virus.**

  - This virus is associated with queen cells which develop dark brown to black cell walls. The cells contain dead pro-pupa or pupa (full of the virus), pale yellow in colour with tough sac like skin (similar to sacbrood).
  - Often noticeable in queenless and broodless colonies being used for cell building with grafted larvae.
  - This viral infection is intimately associated with nosema together with two other viral infections namely: Filamentous virus and Bee virus Y. All three of these viruses multiply only in individual bees that have nosema

5.9 An account of Chilled Brood and its possible causes.

- Chilled brood is brood in all stages which is killed due to low temperatures. All stages means eggs to sealed pupa and for this reason it is very easy to diagnose as no other disease kills brood of all stages in one fell swoop. Bailey states that unsealed larvae can survive several days at room temperatures of c. 65°F, so the temperature drop must be quite severe or prolonged to kill them.

  - Causes:
    - when a colony is approaching starvation,
    - due to spray poisoning (many bees lost),
    - due to mishandling by the beekeeper.
  - Signs:
    - brood of all ages dead,
    - dead brood usually at the periphery of the brood nest,
    - some of the capped cells may be perforated,
    - larvae turn grey to black in colour and remain shiny though they are discoloured,
    - in the later stages a black scale is formed which is easily removable by the bees.
  - Treatment:
    - nil, prevention is better than cure - it is usually the fault of the beekeeper.

5.10 An account of colony starvation and possible remedial actions.

• The beekeeper should never allow a colony to starve, Starvation can occur at any time of the year but more likely in winter, spring and summer due entirely to mismanagement by the beekeeper. Colonies very seldom starve to death if they are left to their own devices. A survey of a random sample of colonies was undertaken in 1977/78 to determine the winter losses. From a total of 1289 colonies, 107 (8·3%) died out and of these 36 (2·8%) were lost due to starvation. If these figures are extrapolated for England and Wales for 1987 with a total of 160,590 colonies, then 4485 are likely to have succumbed due to starvation; a truly disastrous state of affairs.

• Signs of starvation:

   • White pieces of larvae, sucked dry before being thrown out of the hive are the first signs of starvation.
   • The queen may continue to lay and the colony will endeavour to incubate any sealed brood.
   • Due to very reduced carbohydrate intake, the blood sugar levels decrease and mobility is affected. Food sharing continues.
   • The queen is likely to go off lay due to reduced feeding.
   • Drones will be evicted as the food supplies dwindle.
   • Piles of immobile bees are likely to be found on the floorboard either dead or with very reduced mobility.
   • In the final stages the last remaining bees all die together with their heads in the cells and their abdomens protruding.

• Preventive measures:

   • When regular inspections are being undertaken in spring and summer a minimum of 10lb of liquid stores is required to ensure that no starvation occurs up to the next inspection (7 - 10 days hence); this is based on 1 - 1·5 lb per day for a strong colony and is a conservative estimate.
   • Adequate feeding in the autumn to see the colony through the winter to the following April (c. 35 lb of ripened stores).

• Emergency feeding for starvation conditions:

   • Spray warm 50:50 syrup on immobile but living bees.
   • Fill empty cells in brood comb with c.½-1 pt. of the same strength syrup.
   • When bees start to fly, augment with feeder and feed a further 1 gallon of thick syrup (between 50% and 62% sugar).See section 6.12

5.11 An account of the poisoning of honeybees by toxic chemicals and action to take when this occurs and the practical measures possible when prior notification is received.

• The main problems are caused by agricultural spraying of pesticides (a generic name for insecticides, fungicides, herbicides, etc.) for a variety of reasons to combat damage to the crop and hence procure a greater yield for the grower.

- Some growers and farmers undertake the spraying of their own crops, others retain the services of professional spraying organisations. Spraying is a skilled job. Troubles associated with bee fatalities only occur when inexperienced and untrained staff are left unsupervised or the operators take short cuts and/or do not follow the makers instructions. There is now legislation requiring the owner of the crop to provide the beekeeper with a minimum of 48 hours notice if he has colonies nearby and which are likely to be affected (Control of Pesticides Regulations 1986).
- Bees and brood can be killed by toxic chemicals in three ways:

    - By direct contact (through the integument).
    - By eating (into the alimentary tract).
    - By breathing (fumigation into the trachea via spiracles).

- Contact with the poison can occur in three ways:

    - By direct contact on flowers, that the bees are working, which has accidentally been sprayed with the treated crop (eg. weeds in the hedgerows). Note that the treated crop may not necessarily be in bloom.
    - By being caught in the spray on the crop the bees are working.
    - By flying over a crop which is being sprayed.

- Spraying can be done in three ways

    - By fixed wing aircraft (which is the worst - minimum control over the spray).
    - By helicopter (more controlled, note down draught).
    - By tractor (which is the least damaging - ie. to working bees).

- Fruit growers spray the most and cause the least damage to bees. The worst crops for spray damage to bees are crucifers (eg. rape) and field beans. Note that field beans are often sprayed while they are in the early stages of flowering to combat aphid infestation; honeybees are not working the crop at this stage but it is likely to kill off all the bumblebees and therefore ruin a possible honey crop by the honeybees. The authors have found some farmers unaware of the damage they have done in this respect. The time that spraying is actually carried out is very important; this is related to the times that honeybees are expected to be flying:

    - Before 8 am and after 8 pm are the best times.
    - During the day is the worst time irrespective of the weather conditions.

- Diagnosis of spray poisoning:

    - Can be easily confused with CDPV and starvation. Piles of dead bees outside the hive or shivering, staggering and crawling bees also outside the hive.
    - Only laboratory tests and analysis provide a satisfactory answer. Note that a large sample is required because many tests often have to be performed on a large number of small samples to test for the particular pesticide (there are many hundreds of different types). This is the reason for obtaining as much detail as possible about a spray incident, in order to make the identification of the poison as easy as possible.
    - A few or many die quickly which can happen suddenly depending on the poison and how much has been taken in to the colony or how many foragers have been affected.

- The number of foragers at the entrance is less than normal.
- Poisoned bees from inside the hive are ejected.
- Colony tends to become bad tempered.
- The crawling bees tend to have curled up abdomens.
- Returning foragers spin around on the ground until they die.
- Generally there are many dead or dying bees in front of a colony that has suffered poisoning.
- Dead bees usually have their proboscis extended.
- Honey is not usually affected. Poisoned bees are not admitted into the hive and therefore not unloaded by the house bees. Thus, there is no food transfer to other workers providing an automatic protective mechanism.

- Note the recently inaugurated BBKA spray liaison scheme which is understood to be working well in some counties but is lacking in others.
- The most important factor is for the beekeeper to develop a good working liaison with the local farmers and their spray contractors. In the author's experience, the farmers and growers are extremely co-operative and welcome any liaison but the beekeeper has to make the running. If he does, it usually gets to the stage where the farmer or the spray contractor telephones the beekeeper to discuss a spraying operation before it starts.

Action to be taken when spray poisoning is suspected:

- Comply with the agreed procedure of your local spray warning scheme.
- Record as much detail as possible about the incident because if litigation is involved it will be some considerable time in the future.
- Photographs of the colonies and the sprayed crop are often overlooked and are extremely useful at a later date.
- A large sample is required for reasons outlined above; BBKA Advisory Pamphlet # 27 advises 3 samples each of c. 300 bees to be sent to Luddington with the Spray Incident Report which should include the following details:
    - Time and date discovered.
    - # of hives affected plus observations on each.
    - Estimate of dead bees from each hive.
    - Condition of bees and colour of the pollen sample from dead bees.
    - Behaviour of colonies (eg. temper, bees being ejected, etc.).
    - Sketch map of area and OS grid references showing apiary and crop (don't forget to mark North).
    - Weather conditions (wind speed and direction, temperature, rain/fine/sunny /etc.).
    - Discuss with crop owner and seek confirmation of spraying and the spray used.
    - Visit site with owner, if possible, and determine crop acreage and weeds treated.
    - Determine method and time of application together with the flowering state of the crop [nb. photograph of crop].
    - Names, addresses and telephone numbers of all concerned.

- It is important to advise your Branch Sec. and/or your Spray Liaison Scheme representative in order that they may alert other beekeepers in the same area.
- Don't forget to label the samples and mark the hives!

• Practical measures to be taken when prior notice is received:

   • It is now mandatory that 48 hours notice must be given to the beekeeper of any spraying operation.
   • If it is likely that the colonies will suffer then:

      - The colonies should be moved to a safe place if possible; this means to a site at least 3 miles away to ensure that there is no possibility of them returning to the crop.
      - If the colonies cannot be moved, it will be necessary to confine the bees for 24 hours maximum (denying them access to the crop); this is the maximum time that bees can safely be confined providing precautions are taken.
      - The colonies must be prevented from over heating; therefore additional comb should be given together with an empty eke on top with a large sponge soaked in water suspended in it.
      - The colony should be closed up at night.
      - Colonies should then be kept in the dark (tenting with black polythene or covered in straw if available).
      - Colonies can be closed up with crushed ice from a refrigerator; it is claimed that the bees will not pass the entrance even when there are holes in the ice blockade.

   • There is a fair amount of work preparing colonies that are to be closed up for 24 hours and moving them would probably amount to the same effort and be safer particularly if the spray used was lethal for more than 24 hours. In the author's opinion it is safer to move colonies rather than close them, certainly if the colonies are strong and if the weather is warm.

5.12 The expert services available to the beekeeper at national and county level.

• At national level the following organisations can be contacted in the event of information being required on diseases:

   • National Beekeeping Adviser, Luddington, Warwickshire.
      - for analysis of samples for all diseases and poisoning incidents. It should be noted that charges are levied for most of their services and these charges should be ascertained before entering into a contract with MAFF for any services required.
   • IBRA which is now located in Cardiff; an extensive library and many publications for sale are available.
   • BBKA can provide many useful publications and also provides initial advice on legal matters.

• At county level, advice and assistance is available from various organisations depending on which county you reside in:

   • County Beekeeping Association (contact the Secretary).
   • District or Branch Association (usually the first person to contact would be your own Secretary).
   • Your local MAFF offices (to obtain assistance for suspected foul brood infection). It should be noted that most counties only maintain this assistance on a part time basis from April to

October. There is no charge levied for suspected foul brood inspections by these officers.
- County Bee Instructors (CBI) or Lecturers (CBL) will give advice if they are available. Many of the posts are now abandoned and where they do exist many are part time only.
- Agricultural Colleges can often give assistance. Most of the CBIs are based at such a college.

• The secretaries of most district and branch associations are in a position to provide addresses and telephone numbers as they receive (or should receive) all the up to date information.

5.13 The scientific names of the causitive organisms of the bee diseases and infestations.

• Throughout this section the scientific names have been included with the notes on the diseases and infestations concerned. It is suggested that you make your own list now paying particular attention to the spelling of each.

** ** ** **

6.0 BEEKEEPING

6.1 Description of the various types of hive at present in use in the UK.

• Many aspects of practical beekeeping in the UK are complex. If, for example, one considers which is the best hive to use, we are immediately into a complex question because of the wide range of hives available and in use both in the UK and throughout the world. All the hives listed below can be found in use in UK, but it would be incorrect to say that they are widely used. Every beekeeper has to make up his own mind about hives and which he considers best for his own requirements; there are advocates of them all (see 'A case of hives' by L. Heath).

• Below is a list of hives that can be found in the UK, broken down into single and double walled types:

• Double walled hives:

(a) WBC	named after William Broughton Carr	B *
(b) Burgess Perfection	not now produced commercially	B

•   Single walled hives:

(c) British National	first attempt to standardise	B
(d) Modified National	introduced 1960	B *
(e) Smith	developed in Scotland for BS frame	T *
(f) Modified Commercial	attributed to Simmins	B *
(g) Langstroth	most widely used hive in world	T
(h) Dadant	first developed in USA by Dadant	T
(i) Buckfast Dadant	developed by Bro. Adam	B

• Other types:

(j) British deep, Catenary, Long hive, Cottager, Conqueror, etc. together with the skep woven in straw in a variety of shapes and sizes. Copies of some of the standard single walled hives are being made in high density polystyrene.

* these hives may be regarded as the most popular in use in the UK; the Modified National being used in the greatest numbers. B or T = bottom or top bee space. It should be noted that all the single walled hives marked * may be constructed with either top or bottom bee space. The two systems cannot be mixed.

• The essential parts of all the hives are:

• Floorboard and entrance block: the double walled hives (a) & (b) have infinitely variable slides in lieu of an entrance block. The roof should be capable of storing the entrance block.

• Brood chamber: (a), (c), (d) & (e) are all designed for the British Standard (BS) frame with long (1½") lugs except (e) which has short lugs. The other brood chambers are characterised by their different sizes all using larger frames than the BS. The outer dimensions of (d) & (f) are very similar and can be used together particularly the supers of each. Similar to the Smith hive, the Commercial hive uses frames with short lugs, thereby simplifying its construction.

- Supers: the supers for all hives are shallower than the brood boxes. In the case of the Langstroth hive it is common to use the same size boxes throughout, even though shallow supers are available. This has advantages of standardisation but the supers are very much heavier when full, demanding a greater muscle power. Supers for most hives can be equipped with a variety of frames (see section 6.2).

- Crown boards: are usually provided with one or two holes which can be used for ventilation, feeding or fitted with Porter bee escapes may be used as a clearer board. Depending on the type of hive a bee space will be provided on the underside. Generally made of plywood, but can be panelled in glass (a bit of a gimmick for serious beekeeping) which causes condensation on the underside.

- Roofs: these come up in a variety of designs of varying depths from a few inches in the Langstroth to 6" or 9" in other hives. Generally, a deep roof with a minimum clearance between the inside dimensions and the outside of the hive is best in order to prevent it being blown off in windy conditions. All roofs have ventilators, considered by many serious beekeepers to be generally inadequate in size. The standard designs provide wire mesh on the inside allowing them to become blocked from the outside by solitary bees, etc. Putting the mesh on the outside prevents this happening.

- Dummy board: an essential piece of equipment which should be found in every brood chamber and should be considered an integral part of every hive. In our opinion, it is wrong to fill a brood chamber with frames and no dummy board; there should be a dummy board and enough room on its outside face to insert a hive tool to lever the frames tight up together. When the dummy is removed there should be enough room to manipulate the colony without having to remove one end frame.

6.2 Description of the various types of frames in general use in the UK.
[Note: section 6.6 deals with frame spacing. It is very important and is addressed in this section also because of self spacing frames.]

- It is important to thoroughly understand the concept of bee space before considering the pros and cons of various frames available commercially for use in different types of hive [see section 6.3]. Nominal bee space = $\frac{5}{16}$" or 0·312". As all initial work undertaken in UK and USA on frames and bee space was done in imperial units no attempt is made here to use metric equivalents.

- The dimensions associated with frames and their spacing involve a range of fractional imperial measurements which are easy to write in manuscript but difficult and tedious to do on a word processor. For all sizes the following decimal notations will be used:

0·250" = 1 quarter	6 mm
0·312" = 5 sixteenths	8 mm
0·375" = 3 eighths	10 mm
0·437" = 7 sixteenths	11 mm
0·500" = 1 half	13 mm
0·625" = 5 eighths	16 mm

0·813" = 13 sixteenths	21 mm
0·875" = 7 eighths	22 mm
1·063" = 1 and 1 sixteenth	27 mm
1·375" = 1 and 3 eighths	35 mm
1·450" = 1 and 9 twentieths	37 mm
1·500" = 1 and a half	38 mm
1·875" = 1 and 7 eighths	48 mm

• To understand the various types of frame available it is vital to actually see the different types and preferably to handle them; it is an unbelievable jungle and difficult to understand why so many variations are still available for purchase, particularly when many have so little to offer in the way of advantages.

• **Frames general**: they have a top bar, two side bars and two bottom bars. They can have long or short lugs to suit the hive type and they are jointed to allow assembly without glue using only thin nails (gimp pins). They are better assembled using glue and copper pins giving a more robust construction with an infinitely longer life. There are two types of top bar; wedge type and slotted with a saw cut. Both are for fixing wax foundation onto the top bar without recourse to melting wax. The slotted type is a haven for the wax moth to pupate in. Side bars come in two types; parallel sided or with shoulders to provide self spacing (eg.Hoffman), both types usually have a shallow slot as a guide for wax foundation on the inner surface. Bottom bars also come in two types; wide and narrow, the wide type being more robust and discourages the building of brace comb below the frame. All types of frame use wood 0·375" thick for the top and side bars.

• **Brood frames**: top bars are available in two widths namely 1·063" or 0·875" and similarly the side bars; it is possible to mix these with wide top bars and narrow side bars. Spacing of brood frames can be either 1·375" or 1·500" nominal (1·450" = the width of a metal end). When metal or plastic ends are used then the ends of the top bar must be 0·875" wide. Self spacing is generally achieved with Hoffman frames which gives a space of 1·375" between the centre line of each frame. It should be noted that a feral colony building its own comb from scratch, will have a spacing of 1·375" between the centre line of each adjacent comb. If the bees build this way when left to their own devices, one may ask, why use spacers of 1·450" and end up with 1·500", give or take a bit of propolis.

  • Consider now the spaces between top bars and side bars when the two brood chamber spacings are used with the two top and side bar widths.

  • Using 1·375" spacing: with 1·063" frames, the distance between adjacent frames = 0·312"
   with 0·875" frames, ... = 0·500"

  • Using 1·500" spacing: with 1·063" frames, the distance between adjacent frames = 0·437"
   with 0·875" frames, ... = 0·625"

  • Only the first combination meets the criterion of bee space; ie. between 0·250" and 0·375". Using this combination will minimise the building of brace comb between the woodwork of the frames.

• **Super frames**: are usually spaced at 1·500" and 1·875" using metal or plastic ends or 1·625" using Manley self spacing frames. If the wide ends are used they can be staggered to reduce the spacing initially to 1·375" if foundation has to be pulled out. Manley discovered (by experiment) that a spacing of 1·625" was the maximum that can be tolerated for foundation to be drawn satisfactorily and produce thick combs of honey. The Manley frame has the further advantage of the top and two bottom bars being the same width (1·063") thereby providing a guide for uncapping and, of course, there are no ends to remove before extracting. Supers generally require a frame spacing wider than the brood chamber. During a honey flow 'wax builders' are in much evidence and, in the author's experience, unless the bee space concept is observed much brace comb will be built in the supers. If 1·875" spacing is used with 0·875" and 1·063" wide top bars, then the distance between adjacent frames becomes 1·000" and 0·812" respectively; much larger than a bee space!

- **Other frames**: are available but are not in general use, for example:
    - made of plastic instead of wood. They are generally split in two halves in order to equip them with wax foundation.
    - specially made frames for catenary hives. It was expensive to manufacture these to match the catenary of the brood chamber and most users just use a top bar; the bees do not attach comb to the walls of the hive catenary, a most interesting phenomenon.
- **Other types of spacers**: are available but not very popular, for example:
    - plastic Hoffman adapters for converting frames with 0·875" side bars with metal or plastic spacers to self spacing Hoffman.
    - Yorkshire spacers which also attach to 0·875" side bars.
    - Screws or studs in the side bars.
    - Castellated spacers; usually 9, 10 or 11 slots for say a National and should only be used in supers, never in the brood chamber They save removing metal ends before extracting..

6.3 Definition and description of the concept of "bee space".

The Rev. L.L.Langstroth of Philadelphia USA is credited with 'inventing' the bee space in 1851/2. He showed that with a bee space of ⅜" between parts of the inside of the hive, the bees would make little attempt to construct brace and burr comb. He found that by observing this bee space, parts of the hive could be made moveable and interchangeable. This was the turning point from skep beekeeping to modern day beekeeping with the moveable frame hive. The salient points relating to bee space in the moveable frame hive are as follows:

- Bees will propolise a space less than 0·250" and will build brace or burr comb in a space greater than 0·375."

- Bee space is now considered to be 0·312" thereby allowing one sixteenth of an inch tolerance above and below this figure to cater for expansion and contraction of the woodwork while the parts are in use.

- The bee space in a modern hive includes the space between boxes of frames and between the frames and the crown board, the space between the wall of the hive and the side bars of the frames, the space between the walls and the end combs and lastly the space between adjacent top bars and side bars of the frames. All these should be 0·312".

- There is one exception in most hives and that is between the bottoms of the frames in the brood chamber and the floorboard which is of the order of 1". The reasons for this are a bit obscure; in practice the authors have found that during rapid colony build up in the spring much drone comb is built in this area by extending downwards the comb in the brood frames. It also provides a parking space for bees in a large colony in bad weather.

- If a frame spacing of 1·375" is used the space between comb faces becomes 0·500" (ie. two bee spaces of 0·250") allowing the bees to work the two comb faces back to back. If the frame spacing is 1·500" then the inter comb space increases to 0·625" (ie. two bee spaces of 0·312").

- In the brood chamber the only combination of frame and frame spacing dimensions that fully meets the bee space criterion is frames with top and side bars = 1·062" with 1·375" spacing between frames.

- In supers with 0·875" frames and 1·875" spacing between frames, the space becomes one inch between adjacent frames which is much greater than a bee space.

6.4 The purpose of wax foundation within the moveable frame hive.

The purpose of wax foundation is to induce the bees to build straight comb in wooden frames thereby allowing:

- easy inspection of both sides of the comb and facilitating inspection of every cell on the comb face,
- easy extraction of honey,
- easy manipulation of the colony,
- re-use of the wooden frames,
- either worker or drone comb to be constructed as required by the beekeeper.
- minimising the amount of wax that has to be produced by the bees (the bees use the extra wax contained in the extra thickness of foundation compared with natural comb).

The types of wax foundation available are as follows:

- thick foundation, wired in brood chambers and wired or unwired for use in supers,
- thin foundation to be used for cut comb, sections or Cobanas,
- all foundation may be either worker or drone cells,
- sheets are available to fit most types of frame for most types of hive.

The wired foundation is in different styles depending on the manufacturer or supplier as follows:

- the wire can be either straight or crimped,
- the positioning of the wire in the sheet can be either horizontal or diagonal formation (the terminology is a bit misleading - actually a series of V's from top to bottom),
- the material varies from tinned iron, stainless steel to monel metal.

Historical: Mehring in Germany produced the first wax foundation in 1857, not long after Langstroth invented the first moveable comb hive. Weed in the USA produced the first machine to roll foundation in quantity on a commercial basis. Many beekeepers these days make their own foundation as a DIY activity using a simple press or a die between rollers. Foundation for BS brood frames has about 8 sheets per lb. of wax and for Commercial frames about 5 sheets.

6.5 Two methods of wiring frames and embedding wire into foundation.

6.5.1 **General**. Providing wire embedded into the wax foundation gives the finished comb a more stable fixing in the frame rather than relying solely on the adhesion of the comb to the wooden frame. A Commercial brood frame holds c. 7 lbs. of honey when it is full and a BS frame for a Smith or National hive holds c. 5 lbs. With plenty of stores in these frames on a warm day (particularly if the comb is relatively new) it is easy for the comb and adhering bees to part company with the frame and fall to the ground during manipulation if there is no network of supporting wire. It is good beekeeping practice to wire all large size frames ie. those for use in the brood chamber. It is essential to have supers wired if tangential extraction is to be used; if a radial extractor is used, then it is possible to dispense with the wiring but the occasional comb breakage does occur during extraction. The authors never wire supers and always use thin foundation so that the best frames can always be selected for cut comb and those not up to cut comb standard can be extracted.

The wire can be embedded into the foundation and then the wired foundation inserted into the frame. Alternatively the frame is wired and the wax foundation fitted and finally the wire embedded into the wax. All materials expand and contract with a change in temperature and the wire in wax foundation is no exception. In wired foundation the foundation is bent into a slight curve if the temperature is above or below the temperature when it was made. If the frame is wired no change in shape takes place and the foundation and frame remain in the same plane; only the tension in the wire changes providing the frame remains rigid (the timber sizes are adequate for this to be so). The authors preference is for the latter but it is more labour intensive and time consuming than using wired foundation.

Modern wiring is always crimped before it is embedded to minimise movement along the wires and to provide the maximum surface area in contact with the foundation. The size of the wire used is not critical, but it is usual to use SWG 28 to 30. If any part of the wire is exposed at the bottom of a cell, the cell will not be used for brood rearing.

6.5.2 **Two methods of wiring frames**.

(a) The old method uses 4 small hooks attached to the inside surfaces of the side bars 1" to 1½" from the top bar and the same distance from the bottom bars. Call these A and B on one side bar and C and D on the other: A and C being nearest to the top bar. The wire is then made off at A the start point and threaded as follows: AC taught, CA slack, AB, BD slack and looped through CA, DB pulling the whole mesh taught before making off finally at B.

(b) The modern method uses wires parallel to the top and bottom bars. The number of parallel wires depends on the size of the frame, however a distance of about 2" between wires is about right. Holes are drilled in the side bars and small brass eyelets are inserted on the outside faces of them in order to prevent the wire cutting into the woodwork when tensioned. The wire is threaded through the holes and terminated with small tacks after tensioning. The wire is then crimped with a crimping tool (usually two coarse knurled wheels) before the foundation is fitted and the wire embedded.

6.5.3 **Method of wiring foundation and embedding wire**.

This can only be done satisfactorily with a specially constructed wiring board. The sheet of foundation is placed on the board and the wire is then fitted over the foundation by threading it round strategically placed pins to give the required pattern or mesh configuration. Usually half the pins can be levered apart from their opposite numbers to tension the wire before passing a current, for a few of seconds, through the wire to heat it and therefore embed it in the wax foundation. When cool and the wire embedded, the tension is released and the piece of wired foundation removed from the board. The current is usually DC, from a battery charger or 12V battery, of 3 to 5 amps for a few seconds. The timing is by trial and error but dependent on the resistance of the wire which in turn is related to the diameter, length and the material from which it is made.

A similar method is used for embedding the wires of a wired frame into the foundation when it is fitted. Current is passed through the wire to heat it and the foundation is gently pressed on to the warm wires. there is a certain amount of skill required to get the embedding just right but it is easily acquired with a little practice.

A slower method which can be used in both cases is to use a special embedding tool with a spur wheel. This is heated and rolled over the wire on top of the foundation warming both the wire and the wax foundation and pressing it into the foundation at the same time. Again, a certain skill is required for good results.

6.6 The various methods of maintaining the spacing of frames in hives and the measurement of two recognised spacings.

To become familiar with the various types of spacing it is essential to see and handle the bits of equipment involved. There are two major methods of spacing frames. Firstly, self spacing where the dimensions of the frames automatically provide the correct distance between the frames when they are placed in the hive. Secondly, frames can be spaced by attaching specially designed spacers to the frames. Both methods are widely used in the UK.

It is important when manipulating a colony to always ensure that the frames are levered up tight to one side of the brood chamber. Failure to do this each time, will result in a build up of propolis on the spacing surfaces and the spacing gradually becoming too large. Also it prevents, to some extent, uneven combs which are thick at the top with an arch of honey and are consequently difficult to move to a different position in the brood chamber [note that wide top bars and 1·375" spacing obviates this trouble]. It is easier to do the levering with the hive tool between the dummy board and the hive wall. Generally it is the beekeepers who do not use a dummy board that experience this problem; it often gets to the stage where the first frame has to be levered out and forcibly pushed back which is very difficult to do without damaging some of the bees.

6.6.1 **Self spacing**: frames are used in both the brood chamber and in supers. There is one type for brood, the Hoffman frame, where the side bars have specially shaped 'shoulders' with a 'V' on one side and a flat on the other. The overall dimension between the 'V' and the flat is 1·375" giving the same spacing between centres when the frames are placed in the hive. The self spacing frame for supers is the Manley frame which has parallel sided side bars 1·625" wide. These provide 1·625" spacing between the centre lines of the comb when placed in the hive.

6.6.2 **Spacers for attachment to frames**: are listed below with their spacing dimension.

Metal ends
- originally 1·450" but now 1·437" for brood chamber use,
- 1·875" for use in supers.

Plastic ends
- 1·437" for brood frames,
- 2·000" for use in supers.

Double V plastic ends - 1·437" for brood frames (min. contact area).

Hoffman adapters - 1·375" for brood chamber use (2 types available).

Yorkshire spacers - c. 1·500" for brood chamber use.

6.6.3 **Other spacing methods**: include castellated metal strips for use in supers only and screws or studs on the side bar edges. We have found one or two cases where the beekeepers concerned use 'finger' spacing or spacing the frames by eye. This is not recommended except as a temporary measure [nb. it would be very foolish to move a colony without proper spacers].

6.7 An account of the queen excluder and the types in common use.

The purpose of a queen excluder is to exclude the queen from the supers while allowing worker bees access to them thereby keeping all the brood rearing and associated pollen in the lower area of the hive, ie. the brood chamber. Theoretically only honey would be stored in the supers but in practice it is found that the first super often has quite a few cells with pollen stored in them. There doesn't seem to be any answer to this if the brood nest extends close to the top of the brood frames because the bees will always store pollen directly adjacent to the brood where it is required for use. The queen excluder is attributed to Abbé Collins in France in the year 1865.

The general principle is a flat sheet with slotted holes just large enough to allow a worker to pass through [it not only prevents the queen passing through but drones as well]. Zinc sheet is a popular material; the size of the slots was $\frac{5}{32}$" (4mm) or 0·157" but most slotted types are now made with slots of 0·162" or 0·163" depending on which book is read. The same effect can be obtained with a grill of parallel wires.

6.7.1 **Slotted types**: Generally made of zinc but galvanised mild steel slotted excluders are now available. They were made in two versions, one with a series of short slots (c. 1½") and the other with long slots (c. 3"). The design is for bottom bee space hives to allow the flat sheet to lay directly on top of the frames. The long slot variety was easily damaged and the short slot version is generally preferred. It is possible to frame this type of excluder as a DIY job; they are not available commercially in a frame. They are the cheapest of all excluders to buy. The mild steel short slot version is probably the best of those available.

6.7.2 **Wire types**: have all to be constructed with strong rigid wires to prevent damage and bending of the wire during use. The construction must be able to withstand damage by burr and brace comb when it is removed from the hive during manipulations. The gaps between the wires must not be greater than 0·165". All wire type excluders are framed and should have a bee space on one side only; some are on the market with a bee space on both sides which is wrong and they should be avoided. The framed wire type is known as the Waldron excluder and similar types from Germany are known as the Herzog excluder. Both are more expensive than the slotted types. A further type with wood/wire/wood construction is available in USA at an even greater cost; it is claimed the bees 'like it better'(?) than other types. Better ventilation through the hive is achieved with the wire types compared with the slotted types.

6.7.3 Other points of interest:

- If the excluders are electro plated with zinc (and most are) they should be cleaned with boiling water or the careful use of a small blowlamp. They should not be scraped, the plating will be damaged and rusting will occur.

- When replacing an excluder ensure that the top bars are clear of brace comb which may distort the excluder or worse damage it.

- It is quite amazing how often the excluder is put on the wrong way round, the bee space should be below on a bottom bee space hive. Which way should it be placed on the frames; parallel to the frames or at right angles to them? In practice this does not matter as there is a bee space on both sides of the excluder when in use.

- At one branch meeting the authors saw a Waldron type excluder jammed full of dead worker bees stuck in the slots between the wires. It turned out that the beekeeper had been sold an excluder for Apis cerana which is a smaller bee cf. the Apis mellifera - most unusual!

6.8 A detailed account of how to commence beekeeping, including the acquisition of bees, sources of equipment, costs and any precautions necessary when acquiring equipment.

6.8.1 General considerations:

It is essential to obtain some knowledge of bees before actually owning and managing them. This is best acquired by joining the local beekeeping association, attending their meetings both in the winter and summer and by attending any evening classes which may be available.

Reading suitable books, in the early days, is also necessary as well as having a good book readily available at home for reference when required. Two suitable books could be "A guide to bees and honey" by Hooper or "Teach yourself beekeeping" by Vernon; both are relatively inexpensive and readily available. If some of the old MAFF leaflets can be obtained or borrowed, these would also be useful, eg. 283 Advice to Intending beekeepers, 412 Feeding bees, 367 The British National hive and 561 Honeybee brood diseases and disorders. There is a further publication "The beeway code" published by DARG which is a must for all prospective beekeepers.

Before acquiring the bees, a prospective beekeeper must obtain some essential items of equipment; these are: protective clothing (say white overalls), beeproof hat and veil, hive tool and smoker, together with protective gloves and footwear if required. He/she is then in a position to attend association demonstrations and to examine any bees likely to be for sale.

It would be ideal if an experienced beekeeper could take a beginner "under his wing" for a season and personally instruct him/her in his own established apiaries. A year for the beginner without his own bees would make him alert to some of the pitfalls he may likely encounter when setting up on his own.

During this learning process, the prospective beekeeper must make up his own mind on the type of hive and equipment he wants to use. This is a very difficult task for a beginner but it has to be faced and he has to be guided in a non-biassed way. Reading "A case of hives" by Heath could be a good starting point on this decision making process.

The prospective beekeeper has two other decisions to make before acquiring his bees. These are how is he going to start (with a nuc, a full colony or a swarm when available) and where is he to keep them (see section 6.10). It is always best for the beginner to start off in a small way with a nuc early in the season and develop it into a full colony during the season. He should be encouraged to maintain a minimum of two colonies as soon as possible to be able to measure one against the other. This will assist in the overall management, should problems arise with one of them.

6.8.2 **Acquisition of bees**: and possibly the equipment they are delivered or purchased in. The bees can be either a nucleus, a colony or a swarm. A nucleus or a colony can be purchased and acquired at virtually any time of the year except for about three months in the winter, whereas a swarm can only be obtained during the swarming season from say May to August.

The ideal approach must be the acquisition of an overwintered nucleus early in the year (April) after the theoretical learning preparations in the winter. This will allow the new beekeeper to see the nucleus expand and become a colony and to reap a surplus (hopefully) during the first year of beekeeping. The second best method will be the purchase of a full sized colony which is satisfactory if the prospective beekeeper has had some hands on experience. The least attractive is the swarm; stray swarms are a liability mainly because the origin is generally not known and there is often the problem of bad temper as well as possible disease.

Bees can be purchased from the following sources:

- recognised bee suppliers,
- a member of the local association,
- from advertisements in the bee press or local press.

Purchasing bees from a recognised supplier provides some guarantee of quality because of the good name the supplier wishes to retain. The age of the queen should be known and they must be guaranteed disease free; it is not unreasonable to ask for a written assurance on these two points.

If bees are purchased from any other sources it is essential to have the bees inspected by a competent beekeeper and samples taken for disease diagnosis before a deal is struck. A prospective purchaser is strongly recommended to consult the BBKA Standard for Bees, Colonies and Nuclei before making any purchase. It is to be noted that the BBKA define nucleus, colony and stock (colony plus hive). Details can be found in the BBKA year book.

6.8.3 **Acquisition and sources of equipment**: can be considered under three categories, namely, new from recognised equipment suppliers, buying second hand and finally making the equipment as a DIY activity.

Purchasing new: is the most expensive but perhaps the most reliable way of acquiring the necessary equipment. It can be done from the catalogues available and preferably out of season when the rush is not on and discounts may be available. Take advice from experienced beekeepers on what to buy and don't forget to ask why a particular choice is being made (eg. what feeder to purchase and why that particular design).

Purchasing second hand: equipment is always popular but fraught with danger. A very large proportion is rubbish and the newcomer to beekeeping must seek expert assistance in the second hand market. When it has been purchased treat it like the plague and as if every bit is infected with AFB. Every item must be completely sterilised and not brought near the apiary or bees until this has been done.

Constructing equipment at home: as a DIY activity is fine and extremely educational to the newcomer. It is essential that recognised plans are obtained and that the equipment is constructed to standard designs and measurements. If this is not done trouble will arise in the future due to incompatibility.

6.8.4 **Other points for consideration**:

- The cost of new equipment should be obtained from the latest catalogues of the recognised bee equipment suppliers. When buying second hand the new price is of course the yardstick. As a guide, good second hand equipment sells at about a third to a half of the new price.

- Starting beekeeping is an expensive hobby and the prospective beekeeper is advised to get a little manipulative experience before finally taking the plunge on his own. He must ascertain whether he is allergic to bee stings.

- Spare equipment is necessary and advice should be sought on what additional gear to have on standby (eg. for swarm control).

- Finally the newcomers should be encouraged to study for the BBKA Basic Examination.

6.9 The criteria to be observed when moving colonies of bees from one place to another (including optimum distance, vibration, temperature, ventilation and water supply).

6.9.1 **The distance**: that bees can be moved is well known, ie. 3 feet maximum or 3 miles minimum, if no bees are to be lost from the colony concerned. Note that it is usually the stock that is moved not the colony (BBKA definition) because it has to be moved in some receptacle or another. The reason for the distance restriction is twofold. Honeybees forage generally up to a distance of 2½ to 3 miles from their hive and have a 'mental picture' of this area or recognise distinctive landmarks within the area and know how to navigate back using these landmarks. Moving their hive within this known area creates a condition whereby the foragers leave the hive in the new position, re-orientate on leaving the hive but while foraging, recognise well known landmarks and return to the old site. The navigational ability of the honeybee is extremely precise (a matter of a few inches near their own hive). Moving the hive entrance more than 3 feet will create a condition whereby the foragers will not find their hive and will either drift to a nearby hive or cluster at the original position of the hive entrance. The authors conducted a series of experiments some years ago to test the memories of the bees by moving them to a distant apiary and then returning them to a different site in the original apiary. After about two weeks their memories seemed to fail and all foragers returned to the new hive position in the original apiary. For periods away from the base apiary for less than 2 weeks then the bees returned to their original site when brought back. Of course during the 2 week period many of the original foragers would have died a natural death and new foragers would have taken their place. The only time that this is not true is when a swarm issues; it can be hived very close to the original site and the foragers do not return to their original hive. It seems that something very curious happens to their memories (?), rather like erasing a computer disc of all its information.

6.9.2 **Preparing a stock for moving**: starts by removing the crown board and replacing it with a travelling screen, preferably with a space of about 1" on the underside to allow room for any bees to cluster. When being moved, the entrance should be closed (eg. reduced entrance block foam pushed into the reduced entrance just before moving) and not restricted with a screen as many books recommend. If light is showing at their normal entrance they will attempt to escape at this point and there is the danger of them suffocating in the panic to get out. When being moved, the hive parts have to be secured one with another; this can be done in a variety of ways:

- Using a hive strap around everything excluding the roof which is always removed for travelling. Two hive straps in opposite directions are safer than one.

- Screwing plates (4" × 1") at an angle of 45° across the joins between floor and boxes and the screen being fastened with screws to the to box. Note that the 2 plates on each side should be angled in opposite directions to prevent movement. This method is considered to be superior to all others but it is more time consuming. It should be used for a major move over long distances, say greater than 50 miles.

- Spring clips to join the boxes together: these use 3 screws, 2 on one box and 1 on the other.

- Bro. Adam's method of long bolts through the screen, brood chamber and floorboard.

- Using hive staples; these are a bit outdated these days and a fine way of disturbing a colony when hammering them home.

- The entrance block needs securing to the floorboard, the safest way is with two 'L' brackets screwed to the front and to the sides of the floorboard.

Other preparations which are necessary before the actual move are as follows:

- The site and stands at the far end should be ready to receive the stocks immediately on arrival.

- Prepare emergency equipment for journey, ie. veil, smoker, fuel, water spray for occasional cooling, spare ropes, wide sticky tape for accidental bee leaks, etc.

6.9.3 **Moving the stocks**: involves observing some simple rules:

- Place foam in reduced entrance and then remove roof.
- Place the stocks with the frames in a fore and aft direction so that frames cannot swing if emergency braking or stopping is required en route.

- Ensure all stocks are roped down securely before starting. Stop after 15 minutes and check all is secure (tension up if required).

- Corner at slow speed to minimise frames swinging.

- The stocks should be moved preferably during the hours of darkness arriving at the destination about daybreak. If they are moved during the day over heating must be watched carefully and cooling applied (say every hour) with water spray if necessary.

- If they are being moved on a trailer ensure that it has a spare wheel.

- On arrival set up all stocks in final positions replace all roofs and then remove foam at reduced entrances quickly.

- Next day remove screens and replace crown boards.

6.9.4 **Vibration**: excites bees and if they are closed up in transport the temperature increase would be dangerous if insufficient ventilation and cooling were not provided. During transportation by vehicle there will be a continuous vibration keeping the colony in a state of agitation and high temperature. It will therefore be clear that vibration in general is closely allied to temperature and ventilation. In order to minimise these adverse effects, stocks should be handled with care during the loading and off loading process.

6.9.5 **Temperature and ventilation**: go hand in hand and, of course, are allied to vibration. Because of the rise in temperature when a colony is disturbed it is necessary to provide adequate ventilation when moving bees. If very strong colonies are to be moved then it may be advantageous to provide additional space by adding another super as well as providing the ventilation screen. Even with these precautions moving strong stocks during the day in warm weather may be insufficient to prevent dangerous temperature rises, enough to melt wax comb and drown the bees in honey. Spraying the colony with water through the ventilation screen will be required as part of the operation.

6.9.6 **Water supply**: the intent of including this item in the BBKA syllabus under moving bees is not fully understood. Water may be required en route as indicated above and it will be obvious that a regular water supply will be required by the colony when it arrives at its new location.

6.9.7 **Other points**: related to moving bees are:

- Bees should only normally be moved during the flying season, the winter cluster should not be disturbed.

- Continual movement of bees, for say pollination purposes, puts them under stress and stress is the forerunner to nosema.

- It is better to move a stock of bees some days after it has been inspected in order to allow time for the bees to re-propolise all the seals which had been broken. This minimises internal movement of frames etc.

- Travelling screens should be constructed of a mesh of 7 to 1" in a wire gauge of c. 28 SWG.

6.10 The factors to be considered when siting colonies in a small home apiary.

6.10.1 **General considerations**: for siting a home apiary will equally apply to siting colonies in an out apiary. Before the siting of the colonies, consideration should be given first to the site itself and indeed whether it is suitable for keeping bees. Criteria for the site are as follows:

- Is there adequate forage in the surrounding area for the colony to support itself and can it readily obtain water throughout the year?
- There must be no question of danger to humans, particularly children, or animals. One sting can kill an allergic subject if not given expert medical treatment very quickly.
- Ideally, the apiary should be in a place where nobody except the beekeeper can be stung. This criterion is nigh impossible to achieve and siting of the colonies in the apiary becomes of vital importance to minimise this risk.
- Under no circumstances should an apiary be established adjacent to a public thoroughfare, even if there is a barrier (eg.hedge or wall) of suitable height between. Bad tempered bees while being manipulated will attack moving human and animal targets for quite large distances from their hive.
- The site should not be in a frost pocket and protection from the prevailing winds is most desirable.
- The site should be free from any form of flooding or under trees in or on the edge of a wood.
- The site should be accessible by road at all times of the year.
- The site should be surrounded by a stock-proof fence if it is adjacent to pasture where livestock is likely to be grazing.
- Finally a matter of ethics relating to out apiary sites which are often on farmer's land; if another beekeeper has bees close by on the same owner's land, then look elsewhere for another site no matter what the farmer says.

It cannot be over emphasised that the utmost care must be taken in siting an apiary and the colonies in it. Bees cannot be moved around like other livestock. If there is any doubt about any aspect of siting, expert advice should be sought.

6.10.2 **Detailed considerations for siting colonies**: in a small home apiary implies that there is a dwelling house nearby with other people, children and domestic animals, if not in the in the same grounds then on neighbouring property. The following points require attention:

- Stocks must be sited so that the flight path of the bees avoid footpaths and areas where there is likely to be any human or animal activity. Stocks can be sited so that bees have to fly up and over hedges and fences thereby getting the bees to a safe height above anyone on the ground. Under normal circumstances such an arrangement is quite workable but aggressive bees must be considered. It is virtually a necessity to have a 'bolt hole' prepared, over 3 miles away, so that a stock may be moved in an emergency.

- There must be plenty of space around each stock for colony manipulations and maintaining the site (eg. grass cutting). A distance of 6ft. between colonies would not be out of the way for setting up nucs and doing artificial swarms between the adjacent stocks.

- Space should be allowed at the planning stage for expansion in the future, this aspect is often overlooked.

- The layout of the stocks should be in an irregular fashion in order to minimise drifting.

- Hives should be provided with permanent bases to raise the floorboard off the ground to prevent damp and possible rot starting to occur in the lower woodwork of the hive. Concrete bases are undoubtedly the best but try a temporary solution until the site has been tried out for a couple of years.

- The height of the top of the brood chamber to minimise too much bending is very important if large numbers of stocks are involved. Even ½ dozen hives becomes a real pleasure if they are at the right height compared with being too low. A point to consider when designing hive stands.

- In home apiaries it is best to site the stocks out of sight of neighbours if this can be done.

- Bees in the stocks will at some time swarm despite the best efforts of the beekeeper to prevent this happening. Shrubs and trees around the stocks are useful for the swarms to hang on.

- A certain amount of shade from nearby trees is useful particularly at mid-day during the summer. Stocks should not be sited under trees where rain drops can fall from them onto the hive in winter; it disturbs the colony.

Provision should be made for storing spare equipment near to the apiary, preferably a discrete shed for the multitude of bits and pieces. There is a risk to life, albeit fairly remote, wherever bees are kept and if this is remembered when planning an apiary, the chances of success are well assured. Having said all this, we believe that suburban gardens are becoming so small and houses so close together that they are unsuitable sites for keeping bees and certainly not the place for beginners.

6.11 A detailed account of the year's work in the apiary including migratory beekeeping.

The beekeeping year is generally acknowledged as starting after the main crop is removed the previous year. The year's work in the apiary is examined below on a month by month basis starting in August on the assumption that the main flow occurs in July and all the supers have been removed by the beginning of August. Included in the account below are bits of information outside the BBKA syllabus but which all good beekeepers should know.

August: This is the month that work commences to prepare the colonies for winter. For a colony to overwinter successfully the following criteria must be met:

- The colony must be disease free [clustering is the ideal condition for the spread of all diseases].
- A young fertile queen is required to head the colony.
- 35lb. of honey or the equivalent in sugar syrup is required.
- A sound and weatherproof hive.
- Protection from the ingress of mice.
- Some protection from the prevailing winds.

• After supers have been removed, examine the colony and determine the quantity of stores in the brood chamber. This can be estimated frame by frame both by eye and by feeling the weight; an assistant to write down the figures is useful. Take a sample of bees at this inspection to examine for adult bee diseases. If reduced entrance blocks were not inserted when supers were removed, do this at this inspection; robbing is a serious problem at this time of the year when the flow stops fairly abruptly.

• Await the results of disease diagnosis; these results are crucial before starting to prepare the colonies for winter, ie. uniting, re-queening and feeding. Only disease free colonies should be united, never those that are being treated for disease; there is the possibility that the treatment may be unsuccessful.

• Treat for acarine before feeding starts [feeders get in the way and the operation is easier without them].

• Feed all colonies for winter, treating those colonies for nosema as required with Fumidil 'B' in the feed. If after inspection a colony has 15lb. of stores, it will need a further 20lb. of additional stores to see it through the winter; this is equivalent to feeding the colony with 16lb. of sugar [nb. ripe honey contains c. 80% sugar]. It has been recommended by some sources to feed Fumidil 'B' to all colonies as a prophylactic. Many bacteria can develop resistant strains when continually subjected to antibiotic treatment. There is some doubt about this mechanism working in the case of nosema mainly because the pathogen is a protozoa and not a bacterium. Until proof is forthcoming, it would be prudent to err on the side of safety and treat only those colonies with the disease.

• Any necessary re-queening required should be completed before the end of August. The authors consider it best to avoid re-queening at this time if it is possible. If something goes wrong and a colony has to put itself to rights by raising emergency queen cells, there is a dearth of drones at this time of the year for ensuring a successful mating.

September/October: Stocks that have been to the heather will be returning during early September and these will require checking for stores and feeding if required. It is seldom that colonies come back without their brood chambers adequately filled and feeding is a rarity. The final preparations for winter are to be completed and these are done just before the autumn evenings start to develop a chill and a drop in temperature is evident.

• Mouse guards should be fitted early rather than late; mice are also making their winter preparations and seeking a dry warm place to hibernate.

• Ventilation requires attention, c. 4 gallons of water will be produced metabolising 35lb. of stores. Lift the crown board by inserting matchsticks at each corner and close the centre feed hole [the roof ventilators now become inoperative]. Any air flow through the hive will be in through the entrance and out under the crown board and down the sides of the hive under the roof, the smallest opening at either the top or the bottom will control the rate of flow in conjunction with the internal temperature.

• If the green woodpecker is troublesome in the apiary area then protection should be added to the stocks. Polythene can be taped or tied onto the hive sides to deny a foot [or toe] hold to the bird, in which case care should be taken not to interfere with the ventilation. Chicken wire attached to the hive and kept a few inches off it to deny the bird's beak access to the woodwork is also a suitable method of protection.

• The final preparation for winter is the roping down of the stocks if this is considered necessary.

November/December/January: Providing all the preceding preparations have been completed satisfactorily, there is nothing the beekeeper can do to assist his bees to get through the winter to the next spring. Apiaries should be inspected regularly [say once a fortnight] or after a particularly bad spell of weather to ensure all is well. The hives should not be touched, even taking the roof off will raise the temperature of the cluster unnecessarily. If something is amiss then, of course, it must be put to rights; most mishaps seem to be associated with vandalism unfortunately. Heavy snow falls can cover the hive entrances; any snow should be very quietly cleared away without alarming the colony in order to maintain the ventilation at the bottom. The practice of feeding candy on Christmas Day still seems to persist quite widely around the country. It is a nonsense and an unnecessary disturbance to the colony if it has been prepared properly for winter.

February/March: The work to be done will depend very much on the weather and whether the colonies are flying. In the south of England the first warm day after the 3rd week in February is the author's guide.

The first task is to change all the floorboards in the apiary and collect all the scrapings as one large apiary sample and send it away to Luddington for the Varroasis search programme.

• Mouse guards should be removed and reduced entrance blocks put in.

• The last job is to quickly check for sealed stores, remove the matchsticks, lower the crown board and at the same time uncover the feedhole to provide some top ventilation. This all takes a couple of minutes per hive and less with two people; the bees are hardly disturbed.

• When the first task is undertaken on a colony, the hive record for the season should be started and the first entry made. The authors keep two records, one on a hive card kept in the hive roof and the other as a computer printout on a clip board which is up-dated in manuscript in the apiary and then later on the computer. The latest version from the computerised records are taken on the clip board on the next apiary visit.

• In some areas of UK where there is an inadequate supply of early pollen, stocks which are to be used on the rape are fed pollen patties to stimulate brood rearing rather earlier than would have occurred naturally; the patties are put on about the beginning of the month.

• As the weather becomes warmer so the colony will start to fly and forage, with a result, stores are used up at a much greater rate. Water has to be collected to dilute the stored honey [only a 50% sugar solution can be metabolised] and this is a good time to start training the bees to a water supply close at hand.

March: If queens have been overwintered in nuclei this is the time for introduction on a warm sunny day and the bees are flying well.

• This is the first time that the colony is inspected and while removing and caging the old queen, the colony should be assessed and a sample of bees taken for testing for the adult bee diseases. Queens from the over-wintered nucs are then caged after marking and clipping if this is your style of beekeeping; in the author's opinion both are virtually essential. If colour coding is used, a quick method of remembering the colours is as follows:

COLOUR	LAST DIGIT OF YEAR	READ DOWN
W hite	1 or 6	W hich
Y ellow	2 or 7	Y ear
R ed	3 or 8	R eared in
G reen	4 or 9	G reat
B lue	5 or 0	B ritain

The old queens are introduced, in Butler cages, into the nucs to keep them going until queen rearing starts again later in the year. The newly marked and clipped queens are introduced into the colonies also in Butler cages. The whole operation can be completed very quickly and the failure rate at this time of the year is very low. The advantages of this system are:

- While the new queens are in nucs from May to March the characteristics of the queen's protégé can be assessed; only those that are suitable are used.
- Finding queens in March with small colonies is made much easier.
- The nucs are virtually self-supporting and already made up at queen rearing time [something less to do at a busy time].
- Spare queens are available at any time of the year if something does go amiss.

• All the queen cages should be removed during the following two days and a quick check made to see that all is well and the queens are laying in both the nucs and the colonies. Any colonies which are short of stores should be fed.

April: The spring flow will start during the month.

• This is the time that regular colony inspections for swarm control should commence.

• Old comb for replacement is placed at the outer edges of the brood box ready for replacing during the 2nd or 3rd week in the month.

• If any brood chambers are to be changed for repair, maintainance and sterilisation, the colony can be quickly changed to a clean box at this time of the year before supers are required. Supers are added as required above the queen excluders which go on with the first super.

• Supers are added when all except the two outside frames are covered with bees [rule of thumb for both brood box and supers]. Colonies should be building up very quickly and it is better to over super rather than under super early in the season.

• Colonies should be selected for fruit pollination and for going to the rape which will be coming into bloom during April.

• Any colony which is not building up or seriously lagging behind other colonies should be singled out for a special investigation to try and determine the reason. If it can be shown that it is disease free then a re-queening job is more than likely necessary.

May: Usually a very busy beekeeping month with stocks being brought back from the rape and from pollination contracts.

• Regular inspections are continued for swarm control and supers added as required. At every inspection of the colony the following should be checked:

- Are there sufficient stores to last to the next inspection if there is no income available?
- Is the queen present and is she laying normally?
- Is there any sign of disease?
- Is there sufficient comb space for the queen to lay and for the bees [remember many foragers will be out when the colony is inspected]?
- Has the colony built up since the last inspection and/or are there preparations for swarming?

• Towards the end of the month queen rearing should start and arrangements made for making up any nucs that may be required.

• Removal and extraction of the spring crop may also be done during the month and will be necessary if the crop is rape to prevent granulation in the comb. On this point it is necessary to know your area well; although the stocks in the apiary may not have been moved to the rape, it is extremely attractive to bees and they will fly a long way to work it particularly if other sources of nectar are a bit sparse.

• Depending on the weather and colony size the reduced entrance blocks may be removed.

June: The objective is to provide the maximum foraging force and colony size by the end of this month in order to take full advantage of the main flow.

• The colony should still be expanding and further supers may be necessary. This month is notorious in UK for a dearth of nectar and known as the June gap; occasionally it does not happen [eg. in 1989 when nectar continued to flow from March through to end of July].

• If a spring crop has been extracted, colonies may be so short of liquid stores as to require feeding. This requires extreme care to ensure that sugar syrup is not stored in the supers.

July: The main flow usually starts [UK south coast] during the first week of this month and this is what the beekeeper has been preparing for since last August. The colony should be at its peak population just as the flow starts. It is all over by the last week of the month.

• Swarm control inspections are required but it is unlikely more supers will be required for bees [hopefully for honey if the supers are being filled and capped]. With 3 or 4 supers on the colonies it is a hard job lifting them off for swarm control and life is much easier with two people at the job.

• When the flow is complete and the crop ripe then it should be removed and extracted straight away.

• Reduced entrance blocks should be put in to discourage robbing.

• The wet supers should be returned to the hives for drying up after extraction unless it is preferred to store them wet. In a home apiary near neighbours it cannot be over emphasised that wet supers should only be returned to the stocks after dark. Many beekeepers do not understand the reason for this and why literally hundreds of foragers will go milling round the apiary for up to c. 100 metres causing great annoyance to any neighbours in a matter of a few minutes [nb. the round dance and the aroma of honey on the outside of the supers direct from the extracting room]. There is a very great possibility that robbing can be started under these conditions.

• After removing the main crop, any stocks which are scheduled for the heather must be prepared and transported [generally by the end of the month on the south coast]. The essentials for the heather stock are:

- A current year queen [re-queening may be required] to try and keep the brood production going.
- There should be a very full brood chamber with brood on all frames.
- The colony should have plenty of stores to see it through until the heather starts to yield.
- There are mixed opinions on whether the stocks should be manipulated and managed while on the moor [eg. removing brood frames when they are empty to induce greater storage in the supers].
- Drawn comb is generally necessary [usually very cold at 1000ft. altitude] in the supers.

6.12 The principles of feeding a colony of bees.

• The reasons for feeding a colony sugar are shown below:

- To provide adequate stores for winter [rapid feeding].
- To provide emergency stores in the season between colony inspections [rapid feeding].
- As a means of administering drugs [generally rapid feeding].
- To stimulate the queen to lay [usually slow feeding].
- To prevent starvation when the colony is about to succumb [rapid].
- To enhance wax production and the drawing of foundation and comb [slow or rapid depending on circumstances, eg. a swarm on foundation is fed rapidly].
- When a colony has an inadequate foraging force, eg. an artificial swarm which is short of stores [rapid feeding] or after spray poisoning losses.

• The types of feed that are fed to colonies of honeybees:

- The standard feed is white refined household quality sugar either from cane or beet sources [ie. refined sucrose]. No brown or unrefined sugar is permissible.
- It was recommended at one time to feed candy or fondant. It is now used only for special applications [eg. micro mating nucs or the like]. If cream of tartar or vinegar is contained in the recipe, both are toxic to bees cf. refined sucrose. It is best not to feed either candy or fondant if it can be avoided.
- Dry sugar [again refined sucrose] is used by some beekeepers in a tray type crown board usually in the early part of the year supposedly as an insurance policy. It is not recommended because unless water is provided it is extremely difficult for the bees to produce enough saliva to dissolve the crystals.
- Honey. This should only be fed when it comes from the beekeepers own apiary and is known to be disease free. Many imported honeys carry AFB spores and are highly dangerous and must under no circumstances be used.
- Pollen patties are often fed in the early part of the year to provide additional protein where pollen may be in short supply or where colonies are being induced to start brood rearing early. There are two types namely, pollen substitutes [fat free soya flour] and pollen supplements [using trapped pollen; again the source should be from the beekeepers apiaries from disease free colonies].

• The precautions to take when feeding honeybee colonies:

- There should be no spilling or dripping of syrup anywhere in the apiary.
- Precautions should be taken to prevent robbing [reduced entrances and beetight hives].
- Feed should only be administered in the evening just before dark.
- No sugar syrup should find its way into the supers and be mixed eventually with honey for extraction and sale.
- Only pure white refined and granulated sugar should be used.

• Preparing syrup for feeding: Generally there are two types of mix, a thick syrup for autumn feeding which will be stored more or less immediately and thin syrup for spring or stimulative feeding which is to be consumed without storing. Most of the literature quotes the following:

Thick   - 2lb. sugar to 1 pt. of water gives 61·5% sugar concentration
Thin   - 1lb. sugar to 2 pt. of water gives 28·0% sugar concentration
Medium - 1kg. sugar to 1 L of water gives 50·0% sugar concentration

Since the bee requires a concentration of 50% for it to digest and metabolise the sugar then it is clear which is the best one to use if they are to use it straight away. If sugar syrup is to be mixed with cold water, it will be found difficult to obtain a complete mix with 2lb. to 1 pt. The authors use a mix with cold water of 7lb. to 5pt. in an old washing machine [top loader with central agitator]. The concentration works out to be 52·8%, less than 61·5% and hence giving the bees a bit more work to do ripening it to 80% for storing and sealing. As we feed for winter immediately after extracting in August, this causes the bees no distress as they have plenty of time to get their larder in the order they require it before the cold nights set in.

6.13 The most common types of feeders in use.

The requirements of a good feeder are to allow the bees to take the syrup at the rate required by the beekeeper for the management of the colony, while at the same time preventing the bees from drowning in the syrup. Finally, when the feeding is finished, access should be provided for the bees into the feeder so that the bees can clean and dry it up [a job they can do very efficiently given the chance]. There is quite an array of feeders available, not all of them meeting the criteria above and many of them being manufactured in materials that can corrode or are difficult to clean. A further disadvantage of some types is that they are capable of being propolised by the bees so that without maintainance they become unuseable.The various types commonly available are listed below:

• Contact feeders: these come in a variety of shapes and sizes but are all similar in design having a container with a close fitting lid. The lid has a series of small holes or a small piece of gauze through which the bees take the syrup when it is turned upside down over the feed hole or directly onto the frames in the colony. The number of holes regulate the speed that the bees can take the contents. It has the advantage of being cheap and can be readily made at short notice from a bewildering assortment of household containers. The disadvantages are as follows:

- The bees quickly propolise the small feed holes as soon as it is empty.
- As the contents are coming to an end, a change in temperature can force the last remaining content out causing a minor flood of syrup in the hive [usually cleaned up quickly by the bees].
- They are a bit messy to fill and invert without spilling syrup unless one is very careful.
- An eke is required in order to house the feeder under the roof.

• **Round top feeders**: are very widely used in UK and are intended to be placed over a feed hole in the crown board. The capacity varies from c.1 pint to 2 or 3 pints depending on the diameter. The height is usually about 3½ inches. The entry is via a tube in the centre and down the outside of the tube to the syrup. The whole of the centre feeding area is enclosed by a removable cover for cleaning. Older versions were made of metal but now most are manufactured in plastic which is better from a corrosion point of view. This type of feeder is easily filled in situ without the bees escaping in the process. Again an eke is necessary.

• **Miller feeders**: were designed by Dr.C.C.Miller in USA and consist of a tray [about 3" deep] with dimensions in the horizontal plane exactly matching the external sizes of the brood chamber or supers of the hive it is intended to fit. The entry for the bees is via a slot in the centre extending from one side to the other; again it is provided with a cover to prevent the bees from escaping. The capacity is from 1 to 2 gallons. It allows many bees to feed simultaneously thereby allowing very rapid consumption of the syrup [a strong colony can finish the contents in 24hrs.]. Construction is generally in wood with all joints glued to prevent leakage. For bottom bee space hives, a bee space is required on the under side of the feeder.

• **Ashforth feeders**: are virtually identical with the Miller feeder except that the feeding slot is placed at one side allowing the hive to be tilted slightly thereby permitting all the syrup to flow towards the feed slot which is impossible with the Miller type and therefore an improvement. The advantage of allowing all the syrup to be consumed before the tray is opened to the bees for cleaning is that there are no pools of syrup for the bees to drown in.

• **Bro. Adam feeders**: are similar to the Miller and Ashforth except that they have a central entry similar to the Round top type feeders. They are becoming more popular in UK due to some equipment suppliers now manufacturing them. The feeders on the stocks at Buckfast Abbey double up as a crown board (therefore every stock has its own feeder).

All the above type feeders are designed for top feeding. Other feeders are available for internal feeding and bottom feeding [which is seldom practised in UK]. The internal feeder is in the form of a brood frame with wooden sides and an opening at the top to allow access to the bees. The frame feeder is used for feeding nucs; the capacity is inadequate for a colony and few would wish to open the colony in order to feed it. Propolising the float arrangement in the frame feeder to prevent the bees from drowning is a disadvantage of this type.

Other points relating to feeders:

- Each hive should have its own feeder. When feeding starts, particularly in the autumn, all stocks should be fed at the same time.
- There are advantages in combining the feeder as the permanent crown board; it is always available for use and if it stay on one stock it cannot pass on disease by using it on another colony.
- Open tray feeders with straw or polystyrene chips floating in the syrup are often messy and not particularly efficient, the bees often seem to find the 'deep end' and drown in the syrup.
- It is good practice to check the feeders each year for leaks with water before being brought into use.

6.14 The principles of supering.

6.14.1 **Definitions**:

• Super. A box containing frames/combs placed above the brood chamber for the eventual storage of honey. The word 'super' is derived from the Latin word super meaning above [eg. super-script as opposed to sub-script]. Supers are generally shallower than brood chambers because of the weight when full of honey; other than this, there is no technical reason why they shouldn't be any depth providing the frames can be accommodated in the extractor.

• Supering: is the process of adding supers to a colony above the brood chamber either with or without a queen excluder under the super(s).

• Top supering: is the term given to adding further supers to a colony but always adding them on top of any existing supers.

• Bottom supering: no prizes for guessing that the supers are added at the bottom of the pile and always next to the brood chamber.

6.14.2 **Principles involved.**

Reference to the annual colony population cycle graph [and to section 1.11] shows the very rapid increase in adult bee population from the beginning of March. It is not long, providing the weather is fine, that the brood chamber starts to fill up with both brood and bees and if nothing is done there will be insufficient room for the emerging brood. Additional space is therefore provided by adding supers, usually one at a time, as required by the colony build up. On this basis, supers are for bees and, indeed, this can be very true if the colony is using most of or all its income. In such a situation nothing will be stored in the supers and it will be used solely as a parking place for bees in the colony. If this additional space is not provided, overcrowding will occur and this congestion in the hive leads to a breakdown in the food sharing pattern and subsequent distribution of queen substance with a result that the liability to swarm is greatly enhanced.

Reference to section 1.8.7, evaporation, where the manipulation and ripening of nectar to honey was discussed it will be appreciated that large areas of comb are required for the nectar/honey to be 'hung up' to dry in order to evaporate the water. The change in volume of nectar [30% sugar concentration by weight] to honey [80% sugar by weight] is approximately 100:30 thus requiring c. 3.3 times the space for nectar compared with the space required by the finished product.

The calculation looks like this:

1 litre honey weighs c. 1400g   [80% sugar 20% water by weight]
1 litre of water weighs 1000g

1400g honey = 1120g sugar + 280g water
and 1000ml honey = 720ml sugar + 280ml water
∴ 1g sugar has a volume of $720 \div 1120 = 0.64$ml

CONCENTRATION	SUGAR	WATER	TOTAL
Nectar 30%	30g	70g	100g
	19·2ml	70ml	89·2ml
Honey 80%	30g	7·5g	37·5g
	19·2ml	7·5ml	26·7ml

It will be seen that 89·2ml of nectar [30%] when processed to honey only requires a volume of 26·7ml; this is a change of 89·2 ÷ 26·7 = 3·3.

There are only two principles involved as detailed above and summarised below:

- Primarily to provide space for bees, and
- to provide comb area for ripening nectar.

If adequate space is provided for evaporation then it will be clear that there will be adequate space for honey storage.

6.14.3 Other points related to supering:

• By experience it has been found that a good working guide for supering is to add a super when the bees are covering all but the two outside frames of the top box or initially the brood chamber.

• It is better to super early in the spring and be somewhat tardy about adding supers in July when the main flow is on unless this is absolutely necessary.

• In general, top supering is the most widely used method of adding supers. Bottom supering is advantageous if the frames in the super contain only foundation; ie. placing them above the brood chamber, the warmest place in the hive for the wax makers to work.

• There are quite a few beekeepers that super without the use of a queen excluder; however, the majority use an excluder. Again there are beekeepers who advocate not using an excluder in the spring when the first super goes on to encourage the bees into it more quickly. It is true the bees always seem to be somewhat tardy about occupying the first super but this may be due to observing the rule of being just a little ahead of the bees requirements [super when the two outside frames are uncovered].

• See section 6.6 on spacing of frames. Narrow spacing is essential with foundation which can be widened out to 2" when drawn and being filled with honey. The maximum of 2" is the maximum that a colony will build for the storage of honey in the wild state and cut comb containers have been designed on this thickness.

• If the super contains frames with foundation only, it is best to provide one or two frames of drawn comb in the middle to encourage the bees into the super more quickly.

• Wet stored supers are more attractive to the bees in the spring cf. dry supers.

6.15 The importance of supering as a factor in swarm prevention.

The most important factor which causes a colony of bees to swarm is the lack of an adequate threshold level of queen substance throughout the colony which was discovered and proved by a series of experiments by Dr.C.Butler [see sections 1.13, 1.14 and 1.15]. However, it is well known by observation, but not proved, that other factors have an influence. These other factors include:

- season
- shade
- state of flow
- strain of bee
- manipulations

- weather
- ventilation
- district
- comb space [queen]
- comb space [honey]

Considering the two principles of supering, it will be clear that by providing additional empty comb and thereby additional space, not only are the last two conditions relieved but ventilation is also improved. The additional comb space in the supers provides the needed storage space for nectar and honey leaving the comb in the brood chamber for the queen to lay in. The over-riding factor is the prevention of congestion within the hive and the efficient distribution of queen substance.

6.16 A detailed account of one method of swarm control and prevention.

6.16.1 **General**. The subject of swarm prevention [which logically should come first] and swarm control is so vast that to confine the discussion to only one method would defeat the object of having a reasonable understanding of the subject for examination purposes. Firstly, it is necessary to clearly understand the difference between prevention and control and secondly to be able to detect the preparations for swarming and to know how often to inspect the colony in order to detect the preparations.

• **Swarm prevention**: is the action(s) taken by the beekeeper to prevent the colony reaching the state whereby it starts to build queen cells.

• **Swarm control**: is the action(s) taken by the beekeeper to thwart the colony in its endeavours to swarm once the preparations for swarming have been started thereby preventing the loss of bees.

• **Detection of swarming preparations**: this is necessary before any swarm control measures are put into practice. The ability to detect the preparations is prerequisite to any control actions.

• **Frequency of inspections**: for swarm detection is dependent on whether the queen is clipped or unclipped [see appendix 5].

Each will be examined separately. However, a further basic distinction must be made and that is between the stocks in the home [or fixed] apiary and those stocks which have been moved for pollination or to exploit a source of nectar [eg. rape in the spring]. The control method is likely to vary depending on whether the stock is close to hand where additional spare equipment is readily available or on a remote site where spare equipment is not readily available. On this basis two methods of control will be discussed.

It is necessary to control swarming for a number of reasons which are frequently overlooked by many beekeepers, these are:

- A colony that swarms is unlikely to produce a surplus cf. the colony that does not lose its' bees; this is to the detriment of the beekeeper but of little consequence to anyone else.
- Most of the general public are petrified of bees and if not petrified then they have an innate 'api-phobia'. In an urban environment it is essential that no swarm settles on a neighbouring property [this cannot be guaranteed].
- When a colony swarms, there are many thousands of bees flying around which most people find very frightening and can be classed as a nuisance in an urban or suburban environment.

6.16.2 **Swarm prevention**. In sections 1.13 and 1.14 the role of queen substance in relation to swarming was discussed and in section 6.15 the importance of supering as a factor in the prevention of swarming was examined. It is important to understand the role of queen substance and the inter-relationship between food sharing and congestion in the colony as the trigger in the process of swarming. The prerequisite in swarm prevention is that the colony must be headed by a young queen in order that each bee in the colony is assured of its minimum threshold quota of queen substance. When an adequate supply is available at its source [ie. the queen], the next most important factor in swarm prevention is to ensure that the supply can be distributed around the colony; this can only happen if there is plenty of comb for the bees and hence no congestion. Add good hive ventilation and the beekeeper can do little else in the way of prevention. Nevertheless, having done all this a colony may proceed to build queen cells and it is incumbent on the beekeeper to control the issue of a swarm.

6.16.3 **Detection of swarming preparations**. It is very important for every beekeeper to be able to recognise the preparations for swarming while undertaking a routine inspection. At the beginning of the season the colony will have no drones and no queen cups [easily recognised; being almost identical in shape and size to acorn cups]. As the colony builds up, drones will appear and queen cups [known in some parts of the country as play cells; reason unknown] will be built around the outer limits of the brood nest. It is important to examine them closely. If eggs are found in them it does not follow they will be turned into queen cells; in many cases the eggs are eaten by the bees. However, if the cup contains royal jelly, a larva will also be present which is sometimes difficult to see floating in the pool of liquid as the egg may have only recently hatched. This is the sign that preparations for swarming have commenced and swarm control proceedings must be initiated. The simple rules are:

- Dry queen cups [nothing in them or egg only]; the situation can be left to the next inspection.
- Charged queen cups [containing royal jelly]; initiate swarm control procedures.

If all the queen cups have a dull matt finish on the inside, preparations for swarming have definitely not started; the cells will be polished before the queen will lay in them.

6.16.4 **The frequency of inspections** [Reference should be made to appendix A5 - Colony Inspections (timing)].

In order to determine the timing of inspections for swarm preparation it is necessary to understand the mechanism involved and the process of events inside the colony. Subject to the weather being favourable, a swarm will issue with the old queen just after the first queen cell is sealed. If the weather is inclement, then the swarm will not emerge until the weather has improved. The swarm can contain the old queen plus virgin(s) if the time is 8 days after the first queen cell was sealed or rarely virgins only [the old queen having been killed by the virgin(s)] if the time is 8 days or more. Inspection 1 reveals no preparations for swarming and the diagram assumes that immediately the inspection is complete, the colony decides to start preparations to swarm. The process of events is shown for both an unclipped queen [swarm out and lost at sealing of the first queen cell] and a clipped queen [swarm out and back without the queen at the same time]. The colony that had a clipped queen will swarm with the first virgin(s) to emerge at day 16. This all assumes that the beekeeper takes no action after the inspection postulated at day 1.

Inspection 2 reveals unsealed queen cells and the old queen still present at the end of day 6. All the queen cells are now destroyed. It is possible for the colony to build queen cells over 3 day old larvae [the oldest larvae that can become a queen] immediately after the inspection is complete when the old queen is still present. If this happens the diagram shows the process of events if no action is taken by the beekeeper. With an unclipped queen a swarm issues 2 days later and with a clipped queen it issues and is lost in 10 days time.

From the diagram it will be clear that with an unclipped queen inspections must be on a regular 7 day basis [INU - Inspection normal unclipped] and for a clipped queen every 9 days on a regular basis [INC2]. However, with a clipped queen inspections may be on a 14 day basis [INC1] until such time as swarming preparations are found.

6.16.5 **Swarm control**. Over the years there have been three theories of swarming, namely:

- Brood food [postulated by Gerstung in 1890].
- Congestion [postulated by Demuth in 1921].
- Queen substance [postulated and proved by Butler in 1953].

Only the latter satisfactorily explains why a colony swarms and is now accepted as the only correct theory of swarming. Congestion prevents queen substance from being distributed around the colony and is therefore, in itself, not a theory. The brood food theory was accepted for a long time but is now regarded as being incorrect; it is based on the surmise that as the colony builds up, an excess of brood food is produced and this is used in queen cells that are built to absorb this surplus.

Most swarm control methods involve finding the queen and some require finding and destroying queen cells which in turn requires shaking bees off frames. Allied with these operations of controlling, regular inspections are required to know when to undertake them. Such inspections and control can only be undertaken with good tempered bees and ensuring the 'right strain' is a **necessary part** of swarm control. When the colony becomes bad tempered, regular inspections get abandoned, the colony swarms and the bad temper is promulgated further around the district. This indeed must be classed as anti-social behaviour on the part of the beekeeper. The authors believe that a major contributory cause of such situations arising is the present day obsession to wear gloves to manipulate the colony. If the norm were no gloves [kept in reserve for the real emergency] then colonies would be requeened before situations got out of hand. If colonies cannot be handled without gloves, then the handling technique or the strain of bee is at fault and should be corrected without delay. Anyone keeping bees in an urban garden should consider this point long and hard.

Many swarm control methods involve the use of double brood boxes [eg. Snelgrove, Demaree, etc.]; it is not proposed to discuss these here, but to confine the discussion to one method requiring additional equipment [ie. the 'Artificial Swarm'] and the other method requiring only a drawing pin [ie. the 'Destruction of Queen Cells']. If these two methods, suitable for single or double brood box management, are thoroughly understood they can be used for the whole of one·s beekeeping career.

Note that because these two methods are discussed, it does not mean that they are being recommended; there are plenty of other ways of achieving the same end. The destruction of queen cells method requires a clipped queen. As queens have to be found it is infinitely easier to find them if they are marked and marking is regarded as a high priority for effective swarm control.

The Artificial Swarm. This method, which must be common knowledge to anyone keeping bees, is so well known and documented that only a few comments are necessary. Briefly, when the operation has been completed the queen and one frame of bees plus empty comb to fill a new brood chamber remain on the original site and the colony with all the queen cells and remaining bees is put on a new site. All foraging bees return to the original site and, with the queen, form the artificial swarm. The old colony with only house bees and queen cells rear a new queen without swarming. This is the basis of the method, other points of interest are as follows:

- If the colony has supers, then where should these end up; on the artificial swarm with the foragers or on the original stock with the queen cells? Most books show the supers on top of the artificial swarm on the original site. The old stock [now weakened by c.⅓ of the total original number of bees] on a new site may need feeding and could be robbed. It seems logical to put the supers on the old stock and feed the artificial swarm which in all probability will have foundation to pull out and also, doing it this way, there will be no possibility of contaminating the supers with sugar syrup.

- Again many books recommend moving the original stock a second time to draw off any additional foragers 7 days after the manipulation and before a virgin has emerged. Unfortunately the rationale behind such a move is not explained. It does provide additional foragers for the artificial swarm but is not essential to the success of the manipulation.

- It is unnecessary to destroy all but one queen cell in the original stock as the removal of foragers reduces drastically the strength of the colony and the bees will undertake the destruction themselves.

- If necessary, the operation can be completed on the same site with the artificial swarm below and the old stock on top above a swarm board or similar. If it is done this way, then any feeding will be confined normally to the top stock.

- The advantage of this method is that it is virtually 100% successful and can be performed on any stock. Additionally, brood rearing continues with the old queen and the two units can be united at a time suitable to the beekeeper. The disadvantage is that additional equipment is required. As an example, the authors had 8 colonies on the rape in 1988 and all wanted to swarm; to use the artificial swarm method was just not practicable away from our home apiary, and further we did not have enough equipment available at the time. There are horses for courses, and the beekeeper has to make up his own mind how to manage the situation.

The Destruction of Queen Cells. This method thwarts the natural intentions of the bee as opposed to the artificial swarm which complies with their intended actions. Does this have any adverse effect on the colony? There seems to be no straightforward answer to this question because how does one measure the adversity? In the authors experience the colonies appear to work just as well as other colonies but of course there is a break in the brood rearing until the new queen is mated and laying which reduces, to some extent, the honey gathering potential of the colony. Again, reference to the diagram in appendix 5 shows that there are likely to be two conditions at the inspection, ie.

swarming preparations are detected and either the queen is present or she has been lost.

- If the queen is present:

    - Find and cage her [use her in a nuc or destroy later].
    - While the queen is being found, select a good queen cell preferably an open one [sealed ones are sometimes empty!].
    - Mark the frame with the queen cell with a drawing pin [always a good idea to have a spare one inside each hive roof ready for the occasion].
    - Now, and only now, destroy all queen cells except the selected one. The selected frame should be brushed free of bees to ensure that only the one selected cell is left, do not shake it. All other frames must be shaken free of bees otherwise a cell may be missed.
    - 7 days later destroy all QCs except the chosen one, again by shaking every frame except the chosen one which should be carefully handled and the bees brushed off.
    - 21 days later a laying queen should be present [the colony should not be disturbed during this time].
    - With a new queen laying, it is a waste of time doing any more inspections for swarm control during the rest of the season.

- If the queen is not present:

    - If there are no eggs present, this is a sure sign she has gone and is the first thing to ascertain.
    - Select a good queen cell [preferably an open one] but this time, depending on the inspection timing, they may all be sealed and mark the frame as before.
    - Proceed as above at 7 and 21 days.

When swarming preparations are first observed and the queen is present, all frames can be shaken and all QCs destroyed. Sometimes the bees give up the idea of swarming [in the authors' experience in about 25% of the colonies]. As long as the queen is present this can be done a second time. The 3rd time action must be taken.

6.17 Methods of taking and hiving a swarm.

6.17.1 **General considerations about swarms**. Before considering how to take and hive a swarm, a few points of interest are listed below which will assist in understanding the task to be undertaken:

- There are two types of swarm, a prime swarm and a cast. They differ in size, the prime swarm containing about 50% of the original colony and the casts being very much smaller [as little as a cupful of bees in some cases].
- Swarms settle initially within a few metres of the original colony. The prime swarm is generally predictable in its behaviour, remaining where it first settled until it has decided on a new nesting place before moving [a matter of a few hours or rarely a over a week]. In very rare cases they never make up their 'mind' and try to establish comb and a nest outside where they have settled; they invariably perish with the onset of cold and bad weather. Conversely, a cast is very fickle and will take off quickly [can be within the hour] and resettle somewhere else close at hand or at a distance.
- The settling place can be almost anywhere; on a post, on a wall, in a tree or bush, on a fence, under eaves, high or low, etc. Because of this diversity, only broad guidelines can be enunciated for taking them and the beekeeper will have to use his own ingenuity depending on the situation.

- Swarms when they first emerge are generally very docile [even bees of doubtful temper] because they have gorged themselves full of honey before departure for immediate future comb building operations. The longer the swarm hangs up after its emergence, the more its behaviour will return to the normal temperament of the bees and this can be anything from good to aggressive. For this reason swarms from unknown sources should be approached with caution and every effort made to determine their history [eg. how long has the swarm been there?].
- After the swarm has settled it forms a cluster with an outside shell of bees [about 3" thick] with a hollow centre. The outer shell has a small entry exit hole [about 1" diameter]. Close examination of the outer surface will reveal, after about an hour, dancing bees. There may be dances indicating different locations while the swarm is 'arguing' which site to choose as a final resting place.
- Swarms can vary very considerably in size and therefore weight which can range from a few ounces for a cast to 8 or 10 lb. for a prime swarm. The skep or box to transport the swarm must be capable of carrying the load.

6.17.2 **Taking a swarm**. There are different methods depending on the situation of the swarm; these can be classified into the following broad categories:

- Shaking the swarm into a skep.
- Smoking the swarm up into a skep.
- Enticing the swarm into a nuc with a chemical swarm lure.
- Using a frame of brood to attract the swarm onto it.
- Brushing the whole swarm down into a more convenient place so that it can walk into a skep.

The prerequisite of taking any swarm is to get the queen into the skep. Once this is done all the rest of the bees will join her.

The essential equipment required for taking swarms is as follows:

- A good sized skep [mouth to be 14" diameter or larger]. Some swarms are often quite wide and anything smaller makes the operation that much more difficult.
- A second small skep [9-10" diameter]. This is useful for collecting any stragglers if the first shake is not as clean as it might be.
- A piece of cloth or net curtain to close the mouth of the large skep by gathering it up and tying over the top of the skep.
- Secateurs, string and small block of wood to put under the skep [there never seems to be a suitable stone in sight at the right time].
- Butler cage for caging the queen if she is found.

All the above can be kept in the skeps and ready for immediate use [maybe in the back of the car!].Additional items are:

- Smoker, fuel, matches, hive tool and veil.
- Steps and/or ladder.

Many beekeepers take swarms not because they want them but to provide a service to the community. Prompt efficient action to a call is not only appreciated by the person concerned but it is good public relations and enhances the general image of beekeepers.

• **Shaking directly into a skep**. Clear away, with the secateurs, any small foliage to allow the skep to be brought up and under the swarm as close to it as possible. One sharp jerk of the main branch that the swarm is clustering on should get 99% in the skep in one go. The preparation before shaking pays dividends. Slowly turn the skep over and place it in the middle of the sheet on the ground below where the swarm was clustering propped up on one side with a small block or stone to allow the bees to get in and out. In about 20 minutes all the bees will be in the skep and foraging is likely to be starting. Leave until the evening when all the bees have returned, remove the block, gather up the sheet round the skep, tie off and carry away for hiving.

If the queen has been missed in the shaking process the swarm will start coming out of the skep and resettling with the queen, more than likely in the same spot. The small skep is useful now for a second shake if this happens. Throw/shake these bees into the large skep when it is turned over, momentarily, the right way up. It is always best to wait about 20 minutes to see that all is well before departing until the evening. Collecting swarms is much easier with two beekeepers especially if steps or a ladder is involved.

• **Smoking upwards**. Quite often the situation arises where the bees cannot be shaken off [eg. on a wall or a rugged post] and they can then be smoked up into a box or skep. Bees will always walk upwards into a darkened space. When a swarm is on a wall or flat surface this is the only time a cardboard box is better than a skep [a long flat side can be laid against the wall above the swarm]. Remember to push some slivers of wood through the box to support the swarm when it is inside.The box is brought in contact with the swarm and it is gently smoked to get them marching in. Once in they are put on the sheet on the ground as above and left to fly until the evening. Smoking a swarm upwards is a much slower operation than shaking and if the box/skep can be temporarily be fixed in position it will be a lot easier than holding it for half an hour.

• **Using a frame of brood**. This always works in a difficult situation providing the frame can be brought in contact with the swarm. The bees soon cover it and can be shaken into a nuc box and then the frame can go back to the swarm for more bees. If the queen is seen, then the Butler cage will come into its own; with the queen inside the cage and in the nuc box or skep the bees will follow with no prompting. The queen can be released later when the swarm is hived.

• **Chemical swarm lure**. The authors have only tried this once on a cast on a rose bush close to the ground. Two frames of drawn comb were placed in a nuc which was placed with its entrance close to the swarm. The entrance inside the nuc was treated with about ½" of swarm lure [French brand; in a tube to be squeezed out for use like toothpaste]. The idea was to come back in the evening to collect but in minutes the whole swarm was inside the box. Her ladyship, whose garden we were in, thought it magic - and so did we! The experiment seems to have some merit for future use and adaptation for taking swarms in difficult positions.

6.17.3 **Hiving a swarm**. An interesting phenomenon about a swarm is its loss of 'memory' of its old nest or hive more or less immediately it has emerged and settled. The swarm can be taken straight away, hived anywhere in the same apiary and the foragers will not return to the old site. It is analogous to erasing the information on a computer disc. Why such memory erasure should take place during swarming is unknown. Bees have a memory of their original site lasting about 2 weeks when they are not in the swarming mode. There are two basic methods of hiving a swarm:

• Swarm board.
• Shaking into an eke.

The first is fun and amuses both the beekeeper and any spectators. It allows inspection of the swarm and the queen(s) present as they march in. The second method is for the beekeeper who has used the first method so many times that he no longer finds it amusing or if time is at a premium, it is quick and efficient. Swarms are in just the right state for drawing foundation and building comb and such an opportunity should not be missed. If possible always put a swarm onto foundation except maybe for one drawn comb.

• **Swarm board**: is so called because a board [c. 2ft. square] is placed in front of the hive sloping up from the ground to the hive entrance, covered with the sheet from the skep and the swarm shaken on to it. Bees always walk upwards and in a few minutes there is a steady procession walking up into the hive. If the bees are reluctant to start a few taps on the board with a pencil or hive tool will start the proceedings. It takes about ½ hour. The brood chamber that the swarm will occupy should have one drawn comb if the swarm is of unknown origin; if eggs are in this comb 2 days later it indicates an old queen, if there are no eggs it is likely to be a virgin and it should then be left for about two weeks before inspecting again. A feeder with 1 gallon of syrup should be provided straight away.

• **Shaking into an eke**. A shallow eke is placed on the floorboard with the brood chamber complete with frames of foundation over the eke. The entrance should be closed with an entrance block turned through 90°. The brood box is lifted off and the swarm shaken into the eke and the brood box replaced immediately with a feeder with 1 gallon of syrup over. The swarm will walk up into the frames and 10 minutes later the entrance block can be removed [its only purpose was to stop the bees spilling out of the front entrance]. The eke should be removed the next day.

6.17.4 **Other points**.

- Take a sample of the swarm to check for adult bee diseases.
- When brood is starting to be produced examine it very carefully for brood diseases. Swarms from unknown origins can be a liability.
- If any bad temper is present when the colony has settled down, take action to requeen it immediately. Lots of stray swarms are bad tempered bees because their owners could not handle them, so they are a liability on this score also.
- Two or three swarms can all be shaken into one hive at the same time if there is a surfeit of swarms one day as often there is. The queens will sort themselves out and there will be no fighting amongst the bees.
- Casts should be hived with a frame of brood from another colony to prevent them absconding.

6.18 Methods of making nuclei.

6.18.1 Definition of a nucleus.

Before looking at the methods of making a nucleus [popularly referred to as 'nuc' in the singular and 'nucs' for the plural of nuclei] it would be as well to examine the definition of a nuc to understand what has to be made. The BBKA standard [based on the old BS] is that it shall be "a colony occupying not less than three BS [British Standard] combs [14"×8½"] of bees and not greater than five BS combs with the brood [eggs and worker brood] area not less than half the total comb area". As this is a standard for sale it also covers the amount of food and that all the frames should be well covered with bees, etc., etc. The standard for a colony is six BS frames and greater. There seems to be no formally accepted definition of a nuc in past literature but the BBKA standard above will serve our purpose reasonably well as a target to aim at when making a nuc. The number of bees on a well covered BS frame ranges from 1000 to 2000 bees; say an average of 1500 bees [750 on each side]. With 3000 bees it is unlikely that any part of the comb will be seen. It follows that using these figures the number of bees in the minimum sized nuc [3 BS frames] should be about 4500. For the maximum sized nuc [5 BS frames] 7500 bees would be required [c.1·5 lbs in weight]. Nucs on other sized frames would be proportionally sized but with very large sized frames the minimum could not be reasonably less than 2 frames to allow a brood nest temperature to be maintained between the two combs. Other nucs, such as 'mini nucs' and 'micro nucs' are very specialised for queen mating and are beyond the scope of this book. However, their existence should be noted and that they would not fit the definition postulated above.

6.18.2 **The nuc box.**

This is in effect a miniature hive but with some specialised requirements. To consider the principles involved, the definition of a nuc as above will be used. The requirements are as follows:

- It shall be capable of holding 5 BS frames plus a dummy board.
- The inside width to accommodate this shall be $5 \times 1·5" = 7·5"$ plus 0·5" for the dummy board plus 0·25" clearance making a total of 8·25". If Hoffman spacing is used [1·375"] then the clearance will become 0·875". If 1·5" spacing is used it would be preferable to increase the clearance; it is soon used up when inserting a Butler cage for queen introduction.
- A dummy board is essential in a nuc because it will only contain 3 frames when it is initially made.
- The entrance arrangement is important; it should be capable of providing plenty of ventilation but be capable of being restricted to prevent robbing. In built mouse guards are useful if the nucs are used for over wintering new queens for the spring.
- The crown board requires two large ventilators at the back and front covered with wire mesh. A feed hole is necessary to accommodate the nuc's feeder.
- Every nuc box should have its own feeder permanently placed on the crown board. Our preference is for mini Ashforth feeders holding about 1 pint of syrup; they cannot be bought and we have to make our own. If a frame feeder is preferred then the inside dimensions of the box will have to be greater than above; these are not recommended because of opening up the nuc to feed which is bad practice when the nuc is being used for queen mating.
- The roof has to be dimensioned to accommodate the feeder; if the feeder is made too large the roof becomes disproportionate in size and creates too much windage. The most essential features of the roof are that it should be absolutely bee tight [because of feeding] and the ventilators should have an area equal to that of the crown board ventilators. Provision should be made for keeping the record card in the roof so that it is readily available.
- The depth of the box requires a clearance of 1 below the bottom of the frames to accommodate a queen cell protruding from the bottom of a frame. It does not happen very often but it is useful to have.
- The easy mobility of a nuc is essential and each nuc should have its own travelling screen easily and securely fastened when required.

6.18.3 Making the nuc.

The essential components for making a nuc are a queen, bees, food [honey and pollen] and emerging brood. If the nuc is to be used for mating then a QC can be given to the nuc in lieu of a queen. A nuc can be made from:

- a single colony,
- from two colonies or
- several colonies.

There will be a difference depending whether the nuc is to remain in the same apiary or whether it is to be moved more than 3 miles away, the latter being a much easier task.

Method 1. Three frame nuc [to be transported away]. From the parent colony find the queen and cage her. Select two frames of emerging and advanced brood with attendant bees and place in the nuc box. Then select a good frame of food containing fresh pollen and liquid stores, again with bees and also put into the nuc box. Find now a really well covered frame of bees and shake the lot into the nuc. The old queen can be released into the nuc or a new laying queen can be introduced in a Butler cage. The dummy board should be inserted and the space filled on its vacant side with a piece of foam for travelling. Fix the travelling screen and move to the new site immediately. On arrival at the new site open the entrance [reduced] and let the bees fly. If the nuc was destined to receive a QC then the nuc would be transported queenless and without the QC which would be put in at the new site. The nuc should be fed straightaway. Bees will be dying every day through natural causes and these will be replaced by the emerging brood. If a new laying queen is introduced it will be 21 days before any of her protégé hatch out and longer if the QC has to hatch and the virgin to mate before laying commences. During this time the little colony is unbalanced and in a delicate state until it becomes established; therefore, it must be treated with great care to prevent it being robbed. Continual feeding may be necessary.

Method 2. Three frame nuc [to remain in the same apiary]. Proceed as in method 1 but ensure that the frame of liquid stores is virtually full so that the made up nuc can survive without feeding for about 4/5 days. The additional bees shaken into the nuc will be greater in this method to allow for any flying bees returning back to the parent colony. Before shaking into the nuc, lightly shake the frame in the parent colony to get rid of the older bees and then shake the rest into the nuc. Do this with three frames and then introduce the queen and place in a new position in the apiary, out of the flight path of other hives, with a reduced entrance lightly closed with grass. Check after 4/5 days and then feed as required.

If nucs are made up as above taking frames and bees from different colonies, it is prudent to spray each frame lightly with a very weak water and sugar syrup leaving the frames well apart in the nuc box and exposed to the light. The bees that are shaken in should also be lightly sprayed. Finally, slowly bring the frames together after smoking well; it is unlikely that any fighting will occur following this treatment. Alternatively, all the frames less bees can be taken from one or different colonies and the bees from another colony. If it is possible to avoid mixing bees from different colonies, then this should be done. Needless to say, nucs should only be made from disease free colonies.

6.18.4 **Other points**.

- Chalk brood always seems to be a problem with nucs until they become established as well balanced colonies, albeit small ones. The trigger is of course temperature, protein and CO_2 stress. When the nucs are established with good ventilation it has been our experience that they seem able to keep it at bay.
- Nucs made up as 3 frames in early June with QCs, build up to 5 frames and generally collect enough stores to feed themselves for winter. They overwinter well with the young queens and provide the replacement queens for the spring.
- Samples for adult bee diseases should be taken from the nucs and treated in a similar manner to full colonies.
- Nucs should never be allowed to raise a queen from their own emergency QCs; scrub queens will result.

6.19 An account of the various uses to which nuclei can be put.

Nuclei are an essential part of modern apiary management and are probably more useful for teaching purposes than a large full sized colony. Manipulating a large colony can be a daunting experience for the newcomers to beekeeping and their initiation should always be on a nuc. The small colonies should form part of every beekeeping establishment whether it be a commercial honey producing organisation or a small amateur beekeeper with a couple of hives. The number of uses to which nuclei can be put is really quite remarkable, the important ones are listed below:

- Queen mating [these nucs can be quite small, nb. mini nucs].
- Establishing and building into a full colony.
- Increasing stocks and replacing colonies.
- Swarm control.
- Keeping spare queens and breeder queens.
- Assessing the queen's offspring.
- Drawing worker comb.
- Observation hives.
- Requeening large stocks.

• **Queen mating**: probably tops the list of uses and is probably the most complicated. The size can range considerably from the micro nuc with only a few dozen bees to a 5 frame nuc on BS frames. The important feature is that if no brood is present the little colony is prone to abscond. The presence of brood creates conditions which are favourable to the acceptance of a queen cell and there will certainly be no absconding as a mating swarm. Bees will not leave brood which needs tending whether it be young or emerging. Introducing a queen cell to an established nuc always seems to cause confusion with many beekeepers. How long should the nuc be left queenless? There are the following possibilities:

- Remove the queen and introduce a ripe QC [14 days old or greater] straightaway. There is a fair possibility that the cell may be destroyed; a cell protector [or a bit of sellotape] is a good insurance policy.
- Leaving the nuc queenless for about 2 hours gives a high acceptance success rate. Some advocate feeding at the time the cell is introduced but the rationale for doing this is obscure.
- Leave queenless for 7 days and then destroy emergency QCs before introducing the ripe QC. In our experience this is 100% successful.

Making up a nuc specially for mating purposes is the final option. Care must be taken to ensure the brood is only advanced or emerging. If it is left for two days in a queenless condition the ripe QC will be accepted without trouble. Usually 100% successful.

• **Establishing and building into a full sized colony**. This is the ideal way for a beginner to start beekeeping. The ideal time is to take possession of the nuc in March/April and, of course, this will be an overwintered one. It will be capable of being built into a full sized colony and a surplus obtained during the first season. If the nuc is obtained in June with a current year queen, the build up is unlikely to lead to a surplus during the first season.

• **Increasing stocks and replacing colonies**. With the best management and bee husbandry, occasionally stocks are lost during the winter for a variety of reasons. Overwintered nucs are ideal for replacing such losses and will provide a crop during the year of replacement. If stocks are to be increased, then new nucs will have to be made up during the season. The normal time for this is May/June when queens can be also reared and the colonies are strong enough to provide the bees for making the nucs.

• **Swarm control**. Removing bees and brood from strong colonies to make nucs is an effective method of swarm prevention by reducing the colony population. If QCs are present in the parent colony, one of these may be usefully used in the nuc when it is made up. The danger of perpetuating a swarming strain must be taken into consideration with this particular use.

• **Keeping spare queens and breeder queens**. All beekeepers should have a spare queen available for emergency purposes. This means maintaining an overwintered nuc or two in case one is required early in the year when it would be impossible for a virgin to mate due to drones not being available. The life of a breeder queen can be extended by keeping her in a nuc and thereby severely limiting the extent of her egg laying. In fact the genetic material [eggs and larvae] for queen rearing can be obtained directly from the breeder queen in the nuc. Breeder queens can be kept for up to 5 years in this way.

• **Assessing the queen's offspring**. Because of the very widespread problem of bad temper, we consider it essential that new queens are assessed in nucs prior to being introduced into colonies. It is easy to deal with bad tempered bees if they are in small numbers. Other characteristics are observed such as laying pattern, nervousness, amount of propolis collected, etc.

• **Drawing worker comb**. Small colonies produce little, if any, drone comb when compared with large colonies. A nuc will always draw worker comb irrespective of whether foundation has been provided. Therefore, old comb with the drone comb cut out can be given to nucs for repair as well as giving them foundation. All our 5 frame nucs are given 1 or 2 frames of foundation to pull out every year.

• **Observation hives**. These contain only two or three frames of bees and are therefore stocked from a nucleus. A greater use could be made of observation hives for learning and teaching than is done at present. The observation hive can be stocked from a nuc and given starters instead of foundation to observe comb building in progress.

• **Requeening large stocks**. If a queen is purchased or obtained from another source and has been out of the hive for some time, it is best that she is introduced initially to a nuc [there is a better chance of acceptance in a small colony]. When her laying has normalised in the nuc, then she can be introduced into the large colony. For successful queen introduction, it seems that the old and the new queen must be in the same physiological state. An alternative method is to make a nuc from the colony to be requeened, introduce the queen to the nuc and when laying normally the nuc is united with the parent colony, after first removing the old queen, thus bringing it back to its full strength.

6.20 A detailed account of management of nuclei and swarms to turn them into productive colonies.

6.20.1 **General**. Turning nucs and swarms into productive colonies in UK, means bringing them up to maximum strength by the end of June ready for the main flow. Swarms which usually occur from May onwards will never, during the same season, be large enough to gather a crop comparable with a well established colony; there is insufficient time to produce the bees unless special arrangements are made to add brood from other colonies. Similarly with nucs made up about the same time when queens are being reared, the time is too short. The only nuc which can build up naturally to a full sized colony is one that has been overwintered. The productive colony will have a maximum sized brood nest during the first three weeks in June, which will produce the maximum foraging force at the beginning of July. The number of adult bees in the colony will be about double the amount of brood and the amount of brood will be approximately 20 times the daily egg laying rate of the queen. Putting some numbers to this, we have:

Eggs per day	Total brood	Adult bees
1000	20,000	40,000
1500*	30,000	60,000
2000	40,000	80,000!

With this amount of brood, how many BS frames does this represent? The answer to this, of course, depends on the percentage fill in each frame which will not only contain brood but stores of honey and pollen. Again putting some figures to this on the basis of 5000 cells/BS frame, we have:

Stores	Brood	Brood/frame	20k brood	30k brood
25%	75%	3750	5 frames	8 frames
40%	60%*	3000	7 frames	10 frames
50%	50%	2500	8 frames	12 frames

With a reasonably prolific queen and using a 60%* brood/frame [which from experience seems to be about right], then 10 frames of brood will be required in the productive colony. This could be used as a target to achieve in the build up; it will be a massive colony with 60,000 bees. If the brood chamber is full of brood and a flow starts, then the only place nectar and honey can be stored is in the supers, just where it is required.

Colony size is an interesting concept and writers seem to vie with every other to give a higher number [rather like fishing stories!]. Dr. Jeffree, in some of his experiments, measured 510 colonies and the biggest, a really large one in his own words, contained 47,700 bees. So the estimate above is a bit on the large size, but it is interesting to try and quantify it in terms of frames of brood.

Queens do not lay well when there is no income to the colony. Therefore, if there is no flow the colony will require to be fed in order to build it up. A further factor in the equation is whether foundation is to be given or drawn comb; to draw foundation requires a flow or feeding. A flow requires good weather.

We now have all the variables, namely;
- the fecundity of the queen,
- the weather,
- flow or feeding,
- foundation or comb.

6.20.2 Management of the nuc to build it to colony size.

There are three basic ways of managing a nucleus to ensure it is at peak strength to effectively produce a surplus on the main flow, namely:

- brood spreading,
- natural expansion by feeding or on a natural flow,
- augmenting from established colonies.

• **Brood spreading**: is an unnatural disturbance of the brood nest done in such a way as to encourage a more rapid expansion than would have occurred otherwise. It is a highly skilled job and is not recommended for the beekeeper new to the craft. In essence, the frames of the broodnest are re-arranged so that one with a small patch of brood is placed between two with large patches of brood; this then encourages the bees to expand the small patch of brood to the same size as those on the adjacent frames. Full descriptions of the manipulation may be found elsewhere. If it is practised, it is most important to never 'brood split', ie. interposing an empty frame between one frame with brood and the rest of the brood nest. Brood spreading is likely to be undertaken in the early part of the year when temperatures can drop quite suddenly. Under these conditions it is not unusual for brood to be lost due to chilling even in a small colony that has not had its brood nest altered; if brood has recently been spread the situation can be worse. In our opinion, the whole operation of interfering with the brood nest should be avoided if possible.

• **Natural expansion**: of the brood nest, and hence the colony, takes place dependent on the income. The 'colony explosion' that takes place, for example, in a nuc put out onto the rape has to be seen to be believed. The management required in this case is to prevent the storage of honey in the brood chamber. The danger is that the brood nest becomes restricted by a large slab of honey at either end at a time when it could easily continue expanding. The management here is the judicious use of the dummy board and supering over a queen excluder while the brood nest is not yet on its full capacity of frames. This forces the colony to store nectar in the super while allowing only enough spare comb in the brood chamber for the queen to lay in. If there is no flow then no super is required and feeding [slow - a few ounces per day] should be resorted to, enough to keep the colony continually building up without storing too much in the brood chamber and blocking the natural expansion. Care should be taken to evaluate when a natural flow of nectar starts to occur in order to put the first super on. It should be noted that in a good spring and a source of nectar such as rape the queen can start to lay at her peak rate per day and the build up becomes very rapid. If feeding is required, it is important that sugar syrup is not processed by the bees and stored in the supers and subsequently extracted.

• **Augmenting from established colonies**: means providing either brood or brood and bees from other colonies, always providing they are disease free. Care must again be exercised in doing this, the salient points are:

- Since the nuc is a very small unit, only emerging brood should be added and this only one frame at a time. The reason is twofold; first, emerging brood is not as critical to temperature variation as younger brood and second, there is only a limited number of bees in the nuc to incubate it. If it is an overwintered nuc, the little colony will be balanced and there will be only enough nurse bees to deal with its own young brood, any addition will put it under stress.

- Adding bees and brood to a nuc is better, so that the balance of bees to brood is maintained. When a frame with bees is removed from the parent colony, it should be lightly shaken to dislodge the older bees leaving only the young ones on the frame with the brood to be added to the nuc or expanding colony. Old bees are not required as they are foragers and will return to their old site if it is in the same apiary thereby defeating the object of the manipulation.
- Will the added bees fight with the bees in the recipient colony? It is our experience that in the early part of the year when such operations are being undertaken bees do not fight if they are exposed to sunlight for about 5 minutes. Therefore, open up a space [enough for two frames say] where the new comb is to go and put the new frame in the middle of the gap and leave for a few minutes and then slowly bring the frames together and then close the colony up. Alternatively, the bees on the three frames concerned can be sprayed with a very weak syrup before closing up.
- When augmenting, feeding should be avoided because of the possibility of robbing. It therefore follows that there should be adequate stores until the next inspection.

It should be noted that flying bees can be added to a weaker colony by exchanging the positions of the weak and strong stocks in the same apiary. This only provides older foraging bees with a limited life and creates an unbalance between old and young bees in both colonies.

6.20.3 **Swarms**. These are managed in much the same way as the nucs and again the dummy board should be used to prevent storage in the brood chamber.

6.20.4 **The dummy board**: is one of the most useful pieces of equipment when building up colonies. It is also not used as much as it should be mainly because its three main uses are generally not appreciated ie. for providing manipulating space in a full brood box, providing a good surface to lever the full complement of frames together at the end of a brood box manipulation and as described above for temporarily limiting the size of the brood chamber. Every brood box should have its own dummy board; once a beekeeper has learnt the value of these inexpensive bits of equipment, he will never revert to being without them.

6.21 A detailed account of methods of uniting colonies and the precautions to be taken.

6.21.1 **General considerations**. There are various points concerning bee behaviour and beekeeping practice which are of interest before considering the possible methods of actually uniting, these are:

- Both colonies must be disease free; the spread of disease is caused more often by the beekeeper rather than by any other mechanism.
- In beekeeping literature mention will be made of 'colony odour' and 'hive odour'. Butler [of queen substance fame] postulated that colony odour is genetically produced and each colony has its own characteristics. On the other hand, Bro. Adam is of the opinion that there is no such thing as colony odour but that there is a hive odour which depends entirely on the materials of the hive and the income [nectar and pollen] which in turn depends on the weather. The hive odour is carried by the individual bees. No one has disputed the concept of hive odour but it has not been subjected to any scientific experiments or proof.

- During times of dearth there are many guard bees at the entrance of a colony, some of these being potential foragers if forage was available.
- When there is plenty of forage and a flow on, there will be virtually no guards at the entrance. It is likely that all the colonies in the same apiary are working the same crop and the hive odours are likely to be very similar. Under these conditions, drifting bees are accepted in another colony without challenge or fighting.

From the considerations above it is clear that the best time to unite colonies is during a flow. Feeding, particularly with a scented syrup when there is a dearth is the alternative solution although this is not too easy to feed both colonies separately during the uniting process and feeding both separately beforehand is usually the order of the day.

6.21.2 **Methods of uniting**. There are a variety of ways of uniting, the more important methods and variations will be described.These are:

- Newspaper method.
- Direct uniting.

• **Newspaper method**. This method is probably the most widely used and is generally very reliable and successful in use. The principle involved is very simple; a queenright [QR] colony and a queenless [QL] colony are joined together with a sheet of newspaper between them and the bees chew the paper away and intermingle slowly and hence unite. The paper is deposited outside the hive in the course of the next 24 hours.

- The two colonies to be united have to be brought adjacent to one another [see section 6.9] with their entrances in the same direction.
- The manipulation of uniting colonies should be done in the evening when both colonies have virtually finished flying. The reason is obvious, if the bees are flying then some of the returning foragers will be returning to the entrance of a foreign colony and fighting is likely. Once fighting starts more guards are alerted and then all bees from the other colony trying to enter will be involved. This simple precaution is seldom, if ever, recommended in the literature on practical beekeeping.
- The newspaper requires 3 or 4 pin holes made in it to help start the process of paper destruction. This can be done with the corner of the hive tool blade if care is used. It is useful to cover the paper with the queen excluder to stop it blowing around during the manipulation. Note the requirement to remove the queen excluder the following day after uniting to release any drones above it.
- Prior feeding is required if there is no flow on.
- One colony must be dequeened, the first part of the manipulation. Some books have suggested in the past that the two queens will fight it out and the younger queen will succeed. There is no definite proof that this is so and the possibility exists that the surviving queen may be damaged in the fight. Our advice is do your own selection and be sure of the result.
- Now comes the last, but vexed, question of which goes on top and which goes below? There is the QR colony and the QL colony and either may be the strong [STR] one or the weak [WK] one. Consultation of 4 books; recommended reading for the BBKA exams gave the following result:

	BK1	BK2	BK3	BK4
QR or QL on top	QR	QL	QR	--
WK or STR on top	WK	--	STR	WK

The curious thing is that although the authors were recommending a particular approach, not one of them explained why their way was presumably the right way and whether either of the two conditions take preference. If anything preference would be given to the strong colony being above the weak colony on the basis that the weaker would have the minimum guards at the entrance to oppose returning foragers. A case could be made for having the QR colony below with a queen excluder over on the basis that the arrangement can be left for 3 weeks to allow all the brood to hatch in the upper box; the top brood box can then be removed. We are of the opinion that it does not matter which way round they go and any combination will be successful if 3 criteria are observed, namely:

- De-queen one colony.
- Do the manipulation in a flow or with colonies fed for 2 days before the manipulation.
- Do the manipulation just as it is getting dark.

It is not clear where or when the newspaper method originated but it is simple and effective. In some ways it is similar to using a screen between the two colonies for a few days before removing the screen to allow the bees to unite. This method has the disadvantage of requiring a separate entrance for the top colony which is closed when the screen is removed.

• **Direct uniting**. This is usually undertaken with small colonies [eg. 2 nucs to be united] which may, in total when combined, fill a single brood box. The colonies are brought together, one de-queened and again the operation undertaken in the evening as follows:

- Each frame with bees is removed one at a time from the colonies, dusted with flour or sprayed with a weak syrup and placed in the new brood chamber.
- The frames are taken alternately one from one colony and then one from the other and placed in the new brood box also alternately so that there is a complete mixing.
- Care should be taken not to split the brood nests which should combine to make one large one.
- Finally, the bees are heavily smoked and bumped around to create confusion and the colony closed up.

The flour or the syrup gives the bees an immediate job to do and fighting is none existent or else minimised.

The words of Bro. Adam, on the subject of uniting, are of considerable interest; exposure to light has a calming effect on bees and when they have been exposed for some minutes, they will peaceably unite without any other precaution throughout the whole season. We follow his wisdom with small colonies and nuclei but use the newspaper method for larger colonies. A further variation on the direct uniting method is to place the frame with the queen and bees in the new brood box and then shake all the other bees from both colonies in front of the hive, placing the empty frames in the brood box. The shaken bees are sprayed with syrup or dusted with flour and allowed to return to the hive. This method is not one that is recommended these days, the job can be done with less confusion and uproar in the apiary.

6.21.3 **Other points relating to uniting**.

• Swarms can be thrown together [queens and all] into the same hive within a few days of one another without fighting. When there is a surplus of swarms it is a good way of dealing with them.

We have bumped up to 3 swarms all together on the same day and had two brood boxes of foundation drawn out in 2 weeks; when the flow starts such a unit will collect a surplus.

• Some books state that uniting when there are no drones about is a bad time for this operation; the rationale is not understood. Uniting colonies before winter is a classic time for rationalising the apiary.

• It is a well known observation that a strong colony will collect more surplus than two weak ones; it is important to ascertain the reason for weakness. If it is disease or poor queens, then uniting will not alleviate the problem.

• It has been suggested that a colony of laying workers can be united to a QR colony. Such colonies are virtually impossible to requeen *. We disagree that uniting is a solution because if there are laying workers, the colony will have been QL for 3 weeks or more and all the bees will be old ones. If united satisfactorily they will die off quickly by natural causes and the recipient colony will derive little gain from the addition.

      * Recent work in France indicates that it is possible to requeen colonies of laying workers by dipping the queen in a solution of royal jelly [70%] and water [30%] and introducing them directly. The success rate claimed is greater than 70%.

6.22 An account of the methods of dealing with laying workers.

6.22.1 **General**. Pheromones from the queen and from brood inhibit the development of worker ovaries. Therefore, if a colony becomes hopelessly queenless [unable to raise a queen under emergency conditions] some workers will become laying workers; it is the ovaries of younger bees that develop rather than the older ones. As these workers start to lay and brood is produced [albeit drone in worker cells], the pheromone from this brood in turn tends to have a limiting effect on the ovary development and restricts the number of laying workers. By the time stunted drones are being produced, the colony has been without a queen for a considerable time [6 weeks or greater] and most of the bees are old and of foraging age. The colony will have dwindled considerably. Such colonies are notoriously difficult [if not impossible] to requeen by normal methods.

6.22.2 **Methods of dealing with such colonies**.

• The colony is a useless unit and it can be argued that it is not worth wasting time and effort on it. In such a case, the colony is removed about 200 yards from its own site and shaken off the combs onto the ground and the bees allowed to find [beg?] their way into other colonies in the apiary. It is essential that the laying worker colony is disease free if this method is adopted and best carried out in good weather.

• The colony may be united with a strong colony where the population is much greater than the laying workers otherwise there is a possibility that the queen may be killed by the laying workers. As a precautionary measure the queen may be caged for a few days for her own protection. Again the criterion about disease applies.

• As mentioned in para. 6.21.3 above, work during 1989 in France on queen introduction looks extremely promising for colonies of laying workers.

Most experienced beekeepers opt for the first method above and if there is any suspicion of disease the colony is destroyed. Prevention is better than cure in this aspect of beekeeping.

6.23 A simple method of rearing a small number of queens.

6.23.1 **General considerations**. Every beekeeper should rear his own queens, whether he maintains one or two colonies or more. We believe that spare queens should also be available in case of an emergency. With these two thoughts in mind a few criteria can be developed, as follows:

- If spare queens are to be available throughout the year, then they will have to be reared and kept in overwintered nuclei. The emergency may occur in the spring [eg. a drone laying queen is found in one of the stocks] when it is impossible to raise another queen because of the absence of drones at that time of the year for mating.
- One of the biggest problems in UK beekeeping today seems to be the large number of colonies that are bad tempered. If a queen is overwintered in a nuc it is very easy to determine whether her offspring are suitably tempered for use in a large colony. If not, the queen can eventually be culled. Bad tempered nucs are very easy to handle; bad tempered colonies are very difficult to handle and a nuisance to the beekeeper and his neighbours. Queens can be reared from a good tempered line [as they always should be] but due to the mating, a bad tempered strain can result. Mating is out of the control of the beekeeper and he can only monitor the end result. We consider that all queens should first be tested for temper before introduction into the honey producing unit. If all beekeepers followed this advice, their beekeeping would become more enjoyable, neighbours would not be stung and the amount of personal protection could again be reduced to a simple veil.
- Any queen rearing should be planned and this must include consideration of the following:

    - Timing - ready to start at 2nd or 3rd week in May.
    - Selection of the breeder queen.
    - What method of larval transfer is to be used?
    - Selection and preparation of the cell building colony.
    - How many queens are required?
    - Mating nucs.

    Each of the above points will be examined in some detail with a view to evolving a suitable approach for the small hobbyist beekeeper.

- Before setting out to rear any queens it is absolutely essential to fully understand the life cycle and natural history involved from the laying of the egg by the breeder queen to the new queen actually laying [see section 1.1, 1.7, etc.].
- Once the timing schedule has started, there can be no variation; come hail or shine each operation has to be done on time. Opening a colony under an umbrella is not too easy on one's own and in this sort of situation good tempered bees are definitely preferred.

6.23.2 **How many queens?** A method for a small number of queens is required. The wording is very subjective and a small number to some may seem a large number to others. 5 to 10 would, in our opinion, be a small number and 10 to 50 a medium number and 50 to 500 would be in the commercial queen rearing class. The number is important because to rear good queens, the larvae have to be well fed and the more there are the bigger and stronger the cell building colony has to be. There is, of course, a finite limit to the capabilities of the strongest colony in respect of the number of queen cells [QCs] it can successfully raise. Depending on the method used, so the skill and experience of the beekeeper will influence the success rate. This is particularly true if grafting is adopted. It is therefore prudent to allow a safety factor for this 'success rate' and 50% success is quite realistic for the moderately dextrous operator. So if you want 10 aim at 20. It is always best to use simple methods until more demanding techniques are learnt.

6.23.3 **Timing**. The critical factors in queen rearing are:

- Mature drones are required for mating [12 days after emergence].
- Optimum conditions are those associated with the time that colonies swarm naturally.
- A flow is virtually essential and this can be simulated by feeding.
- There must be an abundance of nurse bees for feeding the larvae.
- Royal jelly is synthesised from pollen and this is required in quantity.

Considering all the above factors, they indicate that, in UK, the month of May is generally the time that they all occur naturally. We always aim to have the cell building colony ready by the 2nd or 3rd week in May. If bad weather occurs the start date can slip a few days but once started there can be no slippage. At the first inspection in the spring, the cell building colony should be selected [the breeder is likely to have been chosen the previous season] and this colony must be built up until it is teeming with bees and brood by mid-May. Queen mating nuclei will be required 9/10 days after larvae are introduced into the cell builder and provision for making these must also be included in the timing schedule.

6.23.4 **Selection of the breeder queen**. This is the queen which is selected for her good characteristics which hopefully will be reproduced in her protégé daughter queens. There are many characteristics, but the important ones, in our opinion, for the hobbyist beekeeper are as follows:

- **Temper must come at the top of the list**. It is essential for the breeder queen to come from a colony which produces good tempered bees which can be handled without gloves and with a minimum of smoke. If this requirement cannot be complied with, it will be necessary to buy in a queen of known good temper to breed from. Having ensured the temperament of the breeder queen it does not follow that success is also ensured; the mating of the young virgins finally casts the die.

- **Nervousness**. This trait is exhibited by bees that move quickly and run all over the comb during inspection, finally clustering in a bunch at the bottom of the frame and falling off leaving the comb virtually bare of bees. It is extremely difficult finding a queen in such a colony and they are usually difficult to requeen by normal introduction with a Butler cage. When a frame is removed from a colony during a normal inspection, the bees should remain quiet and completely cover the frame while it is out of the hive. If they can't meet this test, then don't use the queen as a breeder.

- **Swarming**. Do not breed from a strain which swarms prolifically [producing a large number of queen cells]. The Heath bees brought into this country after the I.O.W. disease [Acarine] are inveterate swarmers and the strain is still with us in our mongrel bees. Some years ago we acquired a queen from a fellow beekeeper which passed on the above traits but turned out to be a swarmer when we bred from her; it was impossible to stop them swarming and we had to dispense with them in the end.

- **Disease**. Some strains of bees are more resistant to a particular disease than others. It goes without saying that the breeder queen should come from a colony with a disease free record.

- **Fecundity**. A prolific egg layer means a large colony and lots of bees which in turn means a larger foraging force for honey production. It is to be noted that the very yellow coloured bees [Italians originally] now brought in from New Zealand are very prolific and convert every bit of food into bees. They often require feeding when other strains can survive on their own resources and they tend to be prone to Acarine infection; a pity because they are delightful bees to handle.

After a season has been completed, most beekeepers are well aware of the colony which is headed by a queen with the least undesirable characteristics [unlikely to be the most desirable!] and selection is therefore very easy. One characteristic which is not mentioned above is the tendency to collect propolis in large quantities. With one or two hives it never seems to matter very much but when 20 or more colonies are 'propolisers' it does become a little irksome during manipulations.

6.23.5 **Selection of the cell builder**. The cell building colony will receive the 12 to 36 hour old larvae and build them into queen cells after a period of queenlessness. The colony has to be very strong and teeming with bees and is to be opened at fixed times irrespective of the weather. It is therefore important that this colony should also be of good temper. A further factor for consideration is whether to work a single brood chamber or a double one. With hives using BS frames [eg. British National] a double brood chamber system is likely to be the most popular format for most beekeepers. For the hobbyist beekeeper it is probably more convenient to do his queen rearing at home rather than at his out apiary. In such a case, there could be advantages initially preparing the cell builder at the out apiary and moving it to the home apiary on to a site just previously occupied by another colony so that the cell builder is further reinforced by the flying bees from the stock moved in the home apiary. It should be noted that the cell building colony and the breeder can be one and the same stock if required.

6.23.6 **Larval transfer**. There are a variety of methods of introducing larvae from the breeder queen's colony to the cell builder; some of the more well known approaches are listed below:

- Grafting [something of a misnomer] is physically transferring the larva with a small tool [a grafting tool, another misnomer] from a worker cell in the breeder colony to an artificially made queen cup which is then put into the cell builder.

- Double grafting - a larva is transplanted into a queen cell and then 1 or 2 days later before it is sealed it is replaced with a new young larva.

- Punching out the whole worker cell containing the egg or young larva. The punched out cell is then attached to a suitable bar in a standard frame for insertion into the cell builder. Examples are the Barbeau method and the Stanley method. The advantage is that the actual larva is not touched and therefore reduces any likely damage and hence failure.

- Transfer of a whole frame of eggs and larvae from breeder to cell builder direct. There are variations on this eg. Miller method of cutting the comb in 'Vs', the line of the cut being adjacent to larvae of the right age. A frame can be put 1" over the top of the cell builder horizontally allowing the cells to be formed vertically from the comb face [popularly called the Australian method supposedly because the frame is the wrong way up!]. A more recent approach is to use the Jenter method. Again these methods prevent physical damage to the larvae and require little dexterity as the larvae are selected by the bees.

A simple method of raising a few queens would certainly rule out grafting methods which are essential for raising large numbers of queens and require good eyesight, a steady hand and some experience to have a fair degree of success. When we undertake grafting, we do 10 grafts each to provide a total of 20 in 2 separate rows to check our own dexterity one against the other [so far there has been no clear winner or loser!] which is a useful monitor.

6.23.7 **Mating nucs**. Provision will have to be made for the mating nucs and be part of the overall plan. Because of the vagaries of the UK weather it is probably better to have the nucs made up prior to the day that the ripe queen cells are transferred. If it is pouring with rain the transfer is quite enough to do. See sections 6.18 and 6.19 for details on nucs.

6.23.8 **A simple method**: based on double brood chambers with BS frames and using the breeder colony as the cell builder. 5 to 10 queen cells may be expected. The steps in the process are listed below:

• Prepare the cell building colony for queen rearing, aiming to have the largest population and brood by mid May when there are likely to be mature drones for mating. The stock would be 2 full brood chambers, QEx and 2/3 supers.

• Draw out a programme [like the top part of Appendix 5] with actual dates incorporated below the natural cycle from egg to emergence.

• Re-arrange the colony for queen rearing using a spare empty brood box. Into the spare brood box put all the frames of open brood and eggs and as much sealed brood and pollen as possible but with no queen. The queen and the balance of frames are placed on the floorboard with QEx over. Add the supers and the box of brood with no queen with crown board and spare eke and feeder [1 pint size adequate]. Feed approximately 1 pint per day. As a result of this rearrangement all the nurse bees [or most of them] will join the brood in the top box and in 24 hours queen cells are likely to be started; this must be checked. If not, insert a screen board with entrance under the top brood box. There can now be no possibility of any queen substance reaching the bees in the top box and QCs will be started.

• After 3 days examine the top box and destroy all sealed QCs [these will have been built on larvae 2 or more days old]. Count and leave all other open QCs which should be sealed in another 1 or 2 days time. There are two problems with this system of queen rearing which are as follows:

  • When destroying the sealed QCs most of the bees will have to be brushed off the frames to ensure that all the sealed ones are destroyed. Do not shake your future queens.
  • One can never be quite sure when the remaining open cells are sealed. It is therefore very important to mark up the programme and be ready to remove the ripe QCs on day 14 approximately 2 days before the theoretical emergence. If one or more emerge earlier than your calculated day 16 it will not ruin the whole programme; ensure that they will be safely in their mating nucs and the first one out will not be capable of destroying the lot in the cell builder.

• At this stage nucs can be prepared and left queenless for 4/5 days if they are permanent nucs. This means de-queening them and then destroying all emergency QCs just before introducing a ripe QC on day 14. All frames must be shaken to ensure no scrub QC is left in the nuc.

• On day 14 new nucs can be made up with only sealed brood [no eggs or open brood] and the ripe QCs distributed to them.

• The top box can be left with one ripe QC and when this new queen is mated and laying the top and bottom boxes can be united after removing the old queen in the bottom box. Alternatively, the top box can be used to make 3 good nucs for some of the ripe QCs.

6.23.9 **Other points of interest**.

• Queens are expensive to buy and after travelling they are not in the best condition for introduction. They are only available at the wrong time of the year. The beekeeper with a small number of colonies can produce queens the equal of those at top prices with the material he has available on his own door step if he is so inclined.

• In order to get rid of bad temper it is essential to maintain overwintered nucs to test the new queens from the time they are reared to the following March when they should be introduced. Our own experience with mongrel strains amounts to culling approximately 10% of those reared.

• For 20 stocks we maintain approximately 15 permanent nucs for re-queening purposes. The nucs are virtually self supporting.

6.24 The symptoms of queenlessness and how this may be confirmed.

6.24.1 **The signs [not symptoms]** of queenlessness in a colony can be readily seen by the observant beekeeper both outside the hive and also inside. The beekeeper with only a little experience can generally tell, by looking at the entrance and studying the bees' behaviour patterns, whether the colony is normal or abnormal. It is a sound practice to have 2/3 stocks in the garden and to study the entrances for a few minutes 2 or 3 times a day while the bees are out, both in the summer and in the winter through to spring. With practice it is generally possible to say whether a colony is normal or abnormal by observing the entrance without recourse to examining the interior; this is particularly useful in the early spring as winter is drawing to a close. The art should be developed by all beekeepers. Colonies seldom become queenless of their own accord, it is usually due to an error or series of errors on the part of the beekeeper and his manipulations which cause this condition. A very small percentage of colonies become queenless during the winter due to the death of the queen at a time when the bees do not have the ability to produce a replacement.

6.24.2 **The signs of queenlessness** are as follows:

• If a queen is removed from a colony, within about 10 to 15 minutes, signs of queenlessness are likely to be observed at the entrance. Bees are wandering around the outside of the hive 'looking for' the queen. These bees crawl up the front and sides of the hive and appear to be in an agitated state. Conversely, if the queen is returned to the colony, it takes about the same time for it to return to normality [see section 1.14 and 1.13].

• If the colony has been queenless for some time [say 24hours or more] the foraging will be greatly reduced, there will be apathy among the workers, some of which will be running around somewhat aimlessly.

- The colony will become more aggressive than usual and difficult to handle.

- The first signs inside the colony are no eggs and eventually no brood and the possible appearance of emergency QCs built over worker brood.

- If the colony has no means of re-queening itself it is said to be 'hopeless', for example, when there is only sealed brood or no brood at all in the colony.

- The colony starts to dwindle in size, and if it is left queenless for 3 weeks or more laying workers will start to appear and eggs will be found in an erratic laying pattern. More than one egg per cell is common and drone brood in worker cells start to appear. At this stage the colony has a 'dispirited air' [difficult to describe in words] about it.

6.24.3 **Confirmation of queenlessness**. A colony, of course, should not be allowed to get to the stage of laying workers. Before this stage, a virtually infallible test is to put a frame of eggs and brood of all ages into the queenless colony. If there is no queen [either fertile or virgin] then queen cells will be started on the introduced larvae within 24 hours. If this test is undertaken too late [ie. laying workers present], it may not work. There is only one situation when this test comb may not give a reliable indication and that is after a colony has just swarmed and has a young mated queen. QCs may be built still under the swarming urge. This is not a very common occurrence.

6.25 Description of methods of queen introduction by cage, uniting or direct action, itemising the necessary precautions.

6.25.1 **General considerations**. In order to introduce a new queen it is essential that the colony is queenless; this may be stating the obvious but there are numerous queens lost each year on this count. Young queens are generally nervous and do not behave sedately until they have reached a certain stage of maturity. Bro. Adam believes that this maturity is reached about 8 weeks after the queen starts to lay and it is unwise to introduce a queen until after this stage has been reached. Some of the known facts about queen introduction, mainly empirical and not proved theoretically, are as follows:

- The greater the population in a colony the more difficult it is to requeen.

- Therefore it follows that requeening before April and after August are the best times and this is borne out in practice.

- Requeening in a honey flow is easier than in times of dearth.

- Requeening in the spring has many advantages particularly as finding queens in small colonies is easy and quick.

- Spring requeening is virtually 100% successful.

- Requeening with a queen of the same strain is easier than with a queen of a different strain. Yellow queens into black colonies are particularly difficult.

- Queens should be in the same physiological condition [ie. laying normally] for successful introduction. Queens that have travelled through the post are not in normal laying condition.

- Bro. Adam maintains that 'colony odour' plays no part in the success of queen introduction, it is the behaviour of the queen. If the behaviour of the old and the new queen is the same then it is not an introduction but a substitution.

6.25.2 **Introduction by cage**. The general principle involved is to place the queen into a cage [with a mesh of about 7 or 8 per inch] with some kind of barrier that prevents the queen from escaping but which can be slowly broken down by the bees, thereby releasing the queen. Various types of cage have been designed but most beekeepers these days use the Butler Cage [due to Dr. C. Butler nb. Queen Substance] which is made of wire mesh and plugged at one end with a piece of wood. The dimensions are approx. $4" \times 0.75" \times 0.5"$. In use, a small piece of newspaper is placed over the open end and secured with a small elastic band. It us usual for 2 or 3 pin holes to be made in the paper before placing in the colony. All our own cages have a slight modification by driving a $1.5"$ panel pin through the wooden plug; this leaves $1"$ of pin proud which is a useful fixing device for pushing into the wax comb. Another popular barrier is a plug of candy which is eaten away by the bees [candy = honey with icing sugar mixed in to a stiff paste]; we consider this messy and more complicated than the newspaper. It is important that the mesh is not too small, if it is it will prevent antennal contact with the queen by the workers. Release is generally effected in 12 to 24 hours. The salient points of this method are:

- The queen only should be imprisoned in the cage without any attendant workers.

- The cage should be placed in the brood nest, and if possible, on a frame with brood at all stages of development. The queen should start her normal laying duties virtually as soon as she has been released.

- Remove the cage the following day.

- The method is virtually 100% successful if undertaken in the spring.

- During the active season if a large colony has to be requeened it is more reliable to plug the open end of the cage with wood and leave the queen imprisoned for 2 days. After this time, open up the colony and remove the frame with the queen cage and then remove the wooden plug and let the queen walk out. If she is unmolested after a couple of minutes replace the frame, with her on it, back into the colony. If she is molested and workers start collecting round and over her, put her back in the cage for another 2 days. If a yellow queen is being introduced into a colony of black bees, we leave the queen a full week before supervising the release which should be a sedate and unruffled affair.

The two other cages which are commonly used are:

- The plastic hair curler, used by ladies, suitably plugged at one end and newspaper over the other end.

- The combined travelling/introduction cage, which usually has candy already incorporated in it. They are mostly made of plastic these days whereas in days gone by they were wooden blocks bored out and wire cloth tacked on. The wire cloth is considered unsuitable, the mesh being too fine. It is important to remove the attendant workers before introduction, so that the queen is fed by the receiving colony and not by the attendant workers from her original colony. See 6.25.3 below for queens received by post or other means of delivery.

6.25.3 **Uniting method.** Methods of uniting are described in section 6.21 and cover all the main points of interest. The only other uniting method that requires discussion is for dealing with a queen that has been travelled and out of her colony of origin for a few days. Such a queen is not in a physiological condition for immediate introduction; she must be in lay. If she is to requeen a large colony, the following points should be observed:

- Make up a nucleus from the colony to be requeened and sited adjacent to it ready to receive the new queen. Do not dequeen the colony at this stage.

- Remove the attendant workers and introduce the new queen in a Butler cage to the nucleus. Note controlled and supervised release if considered necessary.

- Once the new queen is accepted allow 1 to 2 weeks for her to start laying normally in the nuc; the longer the better.

- Unite the nuc with the colony immediately it has been dequeened. The usual way for this is by direct introduction back into the space in the brood chamber created when the nuc was made. As a further precaution the new queen can be caged again for a day while the uniting settles down.

- Any uniting should, as stated previously, be done as it is getting dark and the bees have finished flying; it prevents any fighting by guards at the entrance challenging bees from the other colony.

It will now be clear that because queens can only normally be purchased from the suppliers from May onwards, they are only available at a time least favourable to them being introduced to another colony. The moral must be clear, every beekeeper should not only rear a few queens but should always have a spare one on standby.

6.25.4 **Direct introduction.** There are a variety of methods claimed for direct introduction from shaking the dequeened colony out in front of the hive and, in the resulting confusion, running in the new queen with the bees to dunking the queen in a variety of substances [eg. water, honey, etc.]. None are now recognised as modern methods and generally are considered a bit cranky even though some have been advocated by eminent beekeepers [eg. Simmins]. After dunking, the advocates then run the queen in via the entrance or the feed hole at the top. During 1989 there was some serious work undertaken in France on direct introduction by bathing the queen in a mixture of royal jelly and water [70% & 30% respectively]. Trials were carried out on both fertile queens and virgins into nucs, colonies and colonies of laying workers. The success rate was high for all permutations and it is expected that more will be heard of this approach in the future.

6.26 The problem of robbing and methods used to avoid it or to terminate it once it has started.

6.26.1 **General points in relation to robbing.**

- In nature, a concentration of colonies does not occur and therefore robbing is not a problem. It only occurs where the beekeeper has concentrated his stocks on to a single site to form an apiary. The beekeeper with only one stock will seldom have trouble with robbing.
- Robbers are generally bees from another colony but wasps, hornets and ants can also rob a hive.

- Robbing is for honey only, the other hive products such as pollen and propolis attract no attention as plunder.
- Different strains of bees have different propensities to rob other colonies; the Italian yellow strains being the worst, they are inveterate robbers.
- It is more likely to start after a nectar flow has come to an abrupt halt and in times of dearth.
- It is usually started as a result of bad management practices on the part of the beekeeper.
- Robbing can occur between hives in a single apiary or between hives in two apiaries.
- When robbing occurs in an apiary the only method of communication between the bees is by the round dance which only gives information on distance. Because no directional information is available the bees can only search in the near vicinity which may initiate further robbing if a weak colony is discovered.
- It has been suggested, but not proven, that robbers may release a pheromone to mark the site to be robbed.

6.26.2 **Methods to avoid robbing**:

- Prevention is always better than cure, and good apiary practice at all times is usually the answer.
- Because bees are only interested in a free supply of honey/nectar or sugar syrup available in quantity then there should be no spillage or trace of syrup outside any colony or within the apiary.
- There should be no way into any colony except via the designed entrance; all equipment should be bee tight.
- Colony entrances should be adjusted to the size [or strength] of the colony and to the time of the year and flow conditions.
- When there is no nectar flow, colonies should not be kept open for too long during manipulations.

6.26.3 **Methods of detection of robbing**. There are two types of robbing. The first involves fighting at the entrance of the robbed hive and the second is called silent robbing where no fighting takes place at or within the robbed stock. The behaviour of the foraging bees is quite different in the two cases.

Silent robbing: is characterised by the robbed colony continuing to work normally while at the same time the robbers also enter and leave the robbed colony in a normal manner. The robbed colony can itself be robbing another colony at the same time. The only tell-tale sign is the flight of the bees returning directly to another colony in the same apiary.

Robbing with fighting: has two recognisable characteristics. The first is the fighting outside the robbed hive and the second is the flight of the robber bees approach which is nervous and erratic. The erratic zig-zag flight is curious because it alerts the guards of the robbed colony. Once the robber bee alights and is challenged it becomes submissive and often offers food to the guards.
The characteristic common to both types of robbing is the flight of the laden and unladen bee; rear legs forward in the first instance with a full honey sac and with rear legs trailing astern when unladen in the second instance. The normal rear leg position in flight is reversed, ie. a normal forager should not leave the hive full and return empty.

6.26.4 **Methods used to terminate robbing**. There is no effective way to stop robbing the day it starts. Removal of the robbed stock to another apiary is unsatisfactory as it usually gets robbed again at the new site [the colony being possibly marked by pheromone]. The robbing stock is likely to find another weak stock and continue robbing. The following actions are all effective to some degree:

- Remove robbers to a remote site isolated from other colonies in the immediate vicinity.
- Reduce all entrances and make the nucs and weaker stocks a narrow tunnel 2" or 3" long.
- Straw and grass to cover the entrances of both the robbed and robbing hive to confuse both parties has been suggested by some writers.
- Plain glass leant up against the entrance allowing only entrance from the sides.
- Reversal of the robber and robbed colony.

If any signs of robbing do occur, we consider that the first action must be reduced entrances and this is why it is so important to have the hive entrance block always stored in the hive diagonally across the crown board when not in use. Nucs are particularly vulnerable and methods of restricting any nuc entrances immediately must be normal apiary management. Note that many of the equipment suppliers, economising on wood, do not make the trim in the roofs of hives deep enough to take an entrance block - they are very easily rectified by tacking 4 laths to the existing woodwork.

If a robbed colony is moved it is always wise to leave a frame with some stores in it on the site and allow the robbers to clean it out and finish the robbing job to their satisfaction [the one frame can be put in a spare nuc or travelling box].

6.27 Clearing bees from supers.

6.27.1 **General points on clearing bees**.

- By definition, clearing implies a crop has been collected and the flow is over; robbing can easily be started unless care is taken when removing the crop.
- Bees are generally more irritable after the flow and will be more inclined to defend their stores than before the flow finished.
- Entrances must be reduced at the same time that supers are being cleared.
- If more than one super has been used it is common for brace comb to have been built joining one super with the adjacent one, the brace comb being filled with honey. It is virtually essential to remove this brace comb 24 hours before clearing to avoid dripping honey from removed and cleared supers. It is a sticky job to do but well worth having the frames cleaned up and no honey dripping while they are being collected. The brace comb should not be there and emphasises the importance of bee space and the many incorrect frames that are in use. The process known as 'cracking the supers' is not given the attention it needs in modern bee literature. It should be done just before dark. It also prevents the surprise of finding the supers not cleared because of brood in the one next to the brood chamber if the first supers are checked during the cracking process.
- Supers should be removed very early in the morning before the colony has started flying and taken straight to the extracting room for extraction the same day.

6.27.2 **Clearer boards**. There are basically two types one using Porter bee escapes and the other called the Canadian type with long tunnels for the bees to traverse to get from one side of the board to the other.

Porter bee escapes: possibly the most popular device in UK for clearing bees. The following are the salient points about its use:

- The phosphor bronze springs require very delicate adjustment to a gap of 0·063" and to be free of propolis and wax if they are to work satisfactorily.
- Two Porter bee escapes per board should be used for rapid clearing and to ensure that if one escape becomes blocked the other one will still be operative.
- Any clearer board should have an internal bee entrance incorporated in the design with an opening and closing device which can be operated from outside the hive. When wet supers are returned to the hive, the entrance is opened allowing the bees to enter the supers and, conversely, it is closed when they are to be cleared again. It is important that the operating lever allows the roof to be put in place when the supers are off the hive.
- Approximately 24 to 48 hours are required to clear the supers. The time depends very much on the weather and the flying conditions at the time the board is put on, the better the conditions the shorter the time required to clear.
- The bee escapes will require cleaning from time to time. Methylated spirit is an ideal solvent for propolis and wax.

Canadian clearer boards: have the advantage of no moving parts to be propolised and go wrong. The salient points of this mode of clearing is as follows:

- The same time is required [perhaps marginally shorter] to clear; however, if the weather is bad they are not as effective as the Porter bee escape. The bees seem to learn very quickly that they can return to the supers via the same exit route. The supers must be removed at the latest after 48 hours.
- An entrance capable of being opened and closed from outside the hive is required identical to the board with Porter bee escapes.
- If by any chance there may be an odd drone in the supers, they can traverse the exit route without blocking it as would happen with a Porter bee escape.

8 way plastic escape: which is pinned to the underside of the board directly below a suitable hole. There are 8 plastic slots for the bees to reach the brood chamber and there are again no moving parts. The principle is the same as the Canadian clearer board. Our findings are that it works no better than the Canadian board.

6.27.3 **Shake and brush method**. The method appears to be simple and indeed is, if used at the right time.

• A spare empty super is required, to receive the cleared frames, placed on a roof behind the hive [note the roof is not upturned as most books recommend] with a cover cloth over to prevent any flying bees re-entering the cleared frames. The colony is smoked first at the entrance and then at the top to drive the bees downwards in the supers. One frame at a time is shaken free of bees and those remaining on the frame are brushed off with a feather. The frame free of bees is then placed in the empty super. As one super is cleared so that becomes the receptacle for the next and so on.

• If the supers are sealed smoking has little effect on the bees; they are only subdued when they have gorged themselves with honey and when smoked they are not immediately subdued, only driven downwards. After the honey flow has ceased the colony is likely to be more aggressive and will defend their stores. It will be clear that it is not a method to be used by the uninitiated at the wrong time and certainly not in an urban situation.

• Where should all these bees be shaken? We like to shake them back into the hive rather than at the front, only the bees which we brush off land up at the front. The reason for this is that we keep the super covered with a cover cloth except when a frame is being shaken and if the bees are in the hive, we are in control of the situation and not the bees.

• The final consideration is when should this method be used? At a time when the bees are not flying to safeguard against robbing being started; this means early morning or late evening.

6.27.4 **Other clearing methods**. To provide an overall picture other methods of clearing should be noted, these are:

- Mechanical blowers usually powered by electric motor which in turn is powered by a portable generator. Not for the small time beekeeper.
- Chemical repellents. The three most commonly known are:

    - Carbolic acid - not used these days.
    - Butric anhydride - popular in USA.
    - Benzaldehyde [smells of oil of bitter almonds] - used quite extensively in UK and works well. Should be kept in the dark.

6.28 How to prepare and protect colonies for the winter period.

6.28.1 **General.** Preparations for winter should start in August after the main crop has been removed and extracted.There are reasons for this:

- A colony of bees collects all the stores it needs for winter by the end of July under normal circumstances. If all these sealed stores are removed, sugar syrup has to be fed and this also has to be processed, ripened, stored and sealed; this is difficult for the bees to do on chilly days and nights in autumn, particularly the ripening and evaporating the excess water. It is as well to remember that unsealed stores are likely to ferment and fermenting stores are a cause of dysentery.
- All colonies require sampling for the adult bee diseases before the colony settles down for winter. If nosema is present, Fumidil 'B' can be fed with their winter rations; it would be pointless to feed an adequate amount and then find another gallon has to be administered to get the medicament in. If acarine is present the crop has been removed and the colony can be treated without fear of tainting any honey for sale and the treatment [Folbex VA] can be given during good flying weather.
- Colonies may require to be requeened and it is better to know that the new queen is accepted and laying before clustering starts at 57°F.

Those colonies that are destined for the heather are prepared before they go with a young queen and hopefully return with a full brood box of stores and a super of surplus honey.

6.28.2 **Requirements for successful wintering are as follows**:

- A sound and weatherproof hive.
- 35 lb. of liquid stores.
- A young fertile queen.
- The colony to be disease free.
- Good ventilation while excluding mice.
- No disturbance from October to March.

These will be examined to understand the importance attached to each.

6.28.3 **A sound and weatherproof hive.** This item must be self evident but it is quite surprising the tatty quarters some colonies get landed with; roofs in particular seem to be very often inadequate. Double walled hives have roofs that blow off and a secure method of roping or screwing them down is necessary. If a single walled hive roof has the right clearance between the brood box and the inside dimension of the roof [0·312"] it will not blow off; many do not meet this requirement. Weatherproofing means having the hive off the ground on a suitable hive stand so that the floor board is not permanently damp and can dry out when the weather allows it.

6.28.4 **Stores - 35lb. minimum.** The beekeeper who has to feed his bees before March should not be keeping bees; he has not prepared them adequately for their winter hibernation.

- After the crop has been removed every frame has to be examined in August and the stores estimated. This is done by eye and by feel; it is surprising how quickly and expert one can become at this task.It is essential to know how much a full frame weighs [eg. 5lb. for a BS and 7lb. for a Commercial].
- Having totted up the total in the colony, a calculation is required to know how much sugar to feed. Honey contains 80% sugar and 20% water approximately. Suppose the colony has 25lb. of stores, then another 10lb. is required to meet the 35lb. criterion. 10lb. of honey is equivalent to 8lb. of sugar, the amount required to be fed in a suitable solution. It is frightening the number who keep bees and never do this examination and this simple arithmetic.
- See section 6.12 for details on feeding and the strength of syrup to feed.

6.28.5 **A young queen.** By observation over the years beekeepers have come to know that colonies winter better with a young queen compared with an older one. This sort of statement will be found throughout the literature but no explanation of why this should be ever seems to be forthcoming.

• It is unlikely to be related to a quantitative problem of queen substance and the threshold amount available to each bee because the colony naturally reduces in size in the winter thereby allowing more queen substance per bee. Perhaps queen substance has other effects on colony well-being which are yet undiscovered. Young queens are likely to lay better than old queens and this could get the colony away to a better start in the late winter early spring. The other point is what is young; a queen in her 1st, 2nd year etc? Bro. Adam has always maintained that a queen lays better in her 2nd year particularly if she has not been stressed in the first year.

• Requeening in the spring with queens bred the previous year and kept in over wintered nucs will go into winter at approximately 15 months old and carry the colony through winter before being replaced the next March. This system works well. Whether the queen is regarded as old is doubtful, but if she is not replaced at 21 months old, her efficacy thereafter may certainly be expected to taper off quite rapidly.

• Our own fooling in the matter, as a result of experience, is to ensure that queens of 24 months old do not lead a colony into winter if this can be avoided. Good queens for breeding purposes can of course be kept in nucs for much longer periods. Nevertheless, the reasons why young queens are better for wintering do seem obscure and the fact will have to be accepted until a more scientific explanation is forthcoming.

6.28.6 **The colony to be disease free.** It is vital to sample all colonies for adult bee diseases in August so that treatment may be administered if found to be necessary. Before uniting, which is a common occurrence at the end of the active season, the check for disease ensures that no loss originates from either of the two colonies.

Examination for adult bee diseases now costs money if the samples are sent to Luddington, and the price per sample discourages beekeepers to use the service. Many counties have organised their own microscopy service and more individual beekeepers are doing their own [this must be good]. On the other side of the coin more beekeepers are, for example, feeding Fumidil 'B' to all colonies before winter as a prophylactic against nosema. Although there are rumblings that this is unlikely to be detrimental in the long run [development of strains resistant to this antibiotic] it would in our opinion be a wrong course of action until a definitive paper has been prepared on the subject by someone with the right scientific ability.

6.28.7 Good ventilation while excluding mice. A colony during winter, if it metabolises 35lb. of honey, will be required to get rid of approximately 4 gallons of water. This can only be achieved by evaporation. The average rate is 5 pints/month or 3 ozs/day. It is more difficult for evaporation to take place in the damp western side of UK compared with the drier eastern side. These are the facts, the best configuration for achieving this evaporation is still being debated in the bee press and still no one seems to agree on the subject.

• Our own method, which we have used successfully for many years after experimenting with various approaches, is as follows:

> • Heat escaping from the cluster causes the movement of air, warm moist air moves upwards and is replaced by cold dry air at the bottom.
> • All entrance blocks have 9 - 0·375" holes drilled in them spaced equidistant apart across the length of the block. This gives a total cross sectional area of c. 1 square inch to limit the air flow. The block turned through 90° is a normal reduced entrance block with a 4" wide slot. The ⅜" diameter holes form the mouse guard and are 'kinder' to the pollen collectors in the spring.
> • The crown board is raised about ⅛" with matchsticks at each corner. This gives an exit area for air to escape of c. 9 square inches. The feed hole(s) are covered so that the flow of air is round the outside of the cluster and avoiding the chimney effect directly above the cluster. The roof ventilators now play no part in the ventilation system.
> • The mouseguards are put in usually in September before the ivy flow and the crown board is raised as late as possible to stop the gap being propolised. It is interesting to note that if there is no ventilation at the top of the colony, in the spring it is usually a mess with condensation and mouldy outside combs. One associated problem is that some of the stored pollen also develops mould and is then useless to the bees. Strains of bee that produce a lot of propolis will get themselves into this situation if crown boards are raised too early.

6.28.8 No disturbance during the winter period. Once the colony has settled down for winter it should be left undisturbed until the following spring. Experiments have been conducted and it has been found that the cluster temperature is raised quite a considerable amount [up to 10°F] by say just taking the roof off. Such increases in temperature shorten the life of the winter bee and this manifests itself in spring just when the colony requires all the bees it can muster. Hives should never be sited under trees where the drip of water from the branches can cause colony disturbance.

6.28.9 **Other points of interest are**:

- The green woodpecker can spell disaster for a colony if they direct their attention to boring through the side of the hive. They are usually troublesome in very cold weather when they cannot find forage in the hard ground. There are two ways of protection:

    - Surrounding the hive with chicken netting.
    - Covering the hive with a plastic bag [but note this interferes with the ventilation].

- It is desirable that a colony has stores of pollen which can then be used when brood rearing starts after the winter solstice. We have never found this to be a problem but there are probably parts of UK where there is a dearth. The final topping up of pollen stores is during the ivy flow in September/October [nb. winter bees are produced by large pollen consumption].
- Plenty of bees are required for good wintering but making massive colonies by uniting can defeat the object as shown by some experiments done by Dr. Jeffree at Aberdeen University. The old adage that bees do not freeze to death but starve to death is very relevant to the wintering problem.
- The last thing to do is to remove the hive record card from the roof to bring the final years' records up to date and to prepare new cards for the next season.

6.29 The damage caused to colonies by mice and other pests.

6.29.1 **Mice**. These include the common or domestic mouse and the field and wood mouse. They will enter hives in the autumn seeking somewhere dry and warm to build a nest for hibernation purposes. This activity is prompted by the shorter days and a drop in temperature. The moral is to have mouse guards on the hive in plenty of time to ensure that the mouse does not enter.

• Mice feed on pollen, honey and bees. They therefore cause damage to comb, frames and hive equipment. In winter they will disturb the winter cluster and this disturbance can kill the colony if the temperatures are very low. The bees can sting mice to death and they have been known to be embalmed in propolis because the bees cannot eject them from them hive. Any droppings and urine are generally cleared out by the bees.

• Mice have oval skulls and can squeeze through a ⅜" wide slot but they cannot pass through a ⅜" diameter hole. Mice are therefore not a problem to keep out of the hive and if they do enter, it is the fault of the beekeeper not taking the necessary precautions in time.

6.29.2 **Other mammal pests**. These include shrews, rats, moles, squirrels, hedgehogs, etc. All these can disturb an overwintering colony and in this respect can cause damage to it, but many of them are hibernating themselves. We have noticed pronounced scratch marks at the entrance to some of our hives at one apiary and believe it to be due to badgers although we have not caught them in the act.

6.29.3 **Birds as pests**. The main culprit is the green woodpecker in very cold weather. They usually peck through at the hand hold on National and Commercial hives and in a matter of an hour can make a hole of sufficient size to enter. If they are not spotted in time the colony will surely perish for it will occur in periods of hard frost or snow on the ground. Combs, frames and hive parts will be damaged.

• Other birds are swifts, tits, swallows, shrikes, etc. taking bees on the wing [including queens on mating flights]. We have watched sparrows in the early morning sitting on top of the hive waiting for bees to come out, catching them and taking them back to their nest for the fledglings. Pheasants also have a taste for bees; we wondered why one colony at one apiary was very often irritable until one morning we saw a pheasant tapping at the entrance and eating the bees as they came out to investigate.

6.30 Methods of providing a suitable water supply for bees within the apiary.

6.30.1 **Reasons for providing a supply**. Bees will require water if there is no nectar flow in progress; they are forced into a situation where they have to use their own stores. This is particularly so in the spring when the average colony requires about 150gms water/day for diluting stores to 50:50 ratio and in dry hot weather for cooling, this can increase to 1kg/day. It is essential in an urban apiary that bees are not annoying the neighbours by taking water from their swimming pools, ornamental ponds, drains, etc. At out apiaries it is always good beekeeping practice to ensure that a supply of water is available reasonably close to hand; generally this is not a problem in UK and it is not necessary to provide a discrete supply but if drought conditions prevail for any length of time [eg. in 1989] problems could arise.

6.30.2 **Siting of the supply**: is important if the bees are to use it regularly. The main criteria are as follows:

- It should be close to the hives in the apiary.
- It should be in a sheltered spot out of the prevailing winds.
- It should receive maximum sun and be directly illuminated by the sun in the early part of the year when the declination is zero or southerly. Warm water is collected more quickly than cold.

6.30.3 **Design of the supply**. The methods are legion but all should meet two important criteria namely;

- There must be no possibility of the bees drowning and,
- It must never be allowed to dry up or the bees will seek a more reliable source.

The requirement for the supply never drying up implies the provision of an automatic system or self discipline on the part of the beekeeper to regularly top it up. Any automatic system is easily engineered with the use of a ball valve either on a gravity feed system or mains fed.
In order to prevent the bees drowning they should only be allowed to take water indirectly, for example from wet sacking, moss, pebbles or stone chippings, etc.

6.30.4 **Training bees to the water supply**. Any training should start at the beginning of the year in early spring. Adding a small amount of syrup to sweeten the source or putting syrup adjacent to it is a simple way of starting off the process. Once the bees have found another source it is very difficult [impossible ?] to wean them off it in favour of your own supply.

6.31 An account of the management of colonies for the production of sections and cut comb.

6.31.1 **General**. There are 2 types of section, first the traditional square 4¼" × 4¼" made of wood and second the round [or Cobana] 4" diameter configuration originally conceived from plastic drain pipe. The thickness should be as large as possible [c. 1½"]. Cut comb is cut [2⅝" × 3¾"] to fit the clear plastic containers dimensioned to hold approximately 8ozs. The weight of a square section should be approximately 1lb. and a Cobana about 8oz.

• Sections are contained in section racks and each row is separated by a thin metal separator called a fence or divider to prevent the bees joining up the faces of the sections with brace comb. Bee ways are provided in the wooden section and the effect is to provide a series of compartments for the bees to work in, each being the size of a section. The bees are generally reluctant to work this unnatural arrangement and special management is required to produce this sort of honey. Cobanas are filled more quickly [no corners] and the management required is similar to sections.

• Cut comb, on the other hand, is natural comb honey produced in normal super frames and then cut to size when packaged. There are no special management techniques required to produce this sort of honey and it is now very popular in UK at the expense of sections. With equal colonies in the same flow, a greater crop of cut comb can be obtained compared with sections; this, coupled with the easier management and high price that comb honey commands, is the reason for its popularity. There is, nevertheless, a lot of very poor cut comb on sale, being of inadequate thickness, stained by propolis and often not completely capped.

• To produce good quality honey in the comb it is necessary to produce it quickly to avoid staining of wax cappings with propolis and this in turn means that a 'heavy' flow is required. Good comb honey requires very thin foundation to avoid the "fish bone" effect [the crunchy bit in the middle!]. The final requirement, if it is to be produced quickly, is a strong colony.

• With the large amount of rape now being grown in UK, care must be exercised to avoid any contamination of a crop with rape or other brassicas if comb honey is to be produced.

6.31.2 **Management for producing sections**. The salient points in the management of colonies for producing sections are as follows:

- Because the bees are required to work in a lot of small confined spaces it is necessary to force them into the sections.
- The only way of doing this, is to create congestion in the colony so that the only place that the bees can go is into the sections.
- Creating this congestion disrupts the process of food transfer and the normal distribution of queen substance with the result that the colony is very likely to move into the swarming mode.
- A very young queen is required preferably only a month or two old.
- If such a young queen is used then sections can only be produced on the main flow in July.
- Because of the very real problem of swarming, a strain of bee that is not inclined to swarm is most desirable.
- Normal colony management will be to create maximum colony population by the end of June to take advantage of any main flow which occurs usually in the first weeks of July. The colony should be artificially congested just as the flow is starting by either:

— 135 —

- removing the existing supers [and their attendant bee room] and replacing with a single rack of sections or
- outside frames could be removed from the brood chamber and this area restricted with dummy boards. A judicious mixture of both courses of action could also be applied.

- With a double brood chamber system of management the colony is easily reduced to one brood chamber by rearranging the frames with brood into one chamber.
- Inspection for swarm control is vital if the colony is congested.
- If a good flow does not materialise or the colony persists in its intention to swarm it is often better to give up the idea of sections and revert back to normal.
- The woodwork of sections should be treated with paraffin wax before being fitted into the section rack; this makes them much easier to clean [removal of propolis] before packing for sale.
- Some beekeepers fit the sections with a starter only rather than complete with foundation. This is a matter of preference. With wax starters the bees often produce drone comb but the mid-rib will be thin. With foundation, worker comb is ensured which looks better as a finished product.

If the beekeeper only wants a few sections for his own table, a few hanging sections can usefully be placed in the normal supers next to the brood chamber just when the flow starts. The hanging section is a special frame that holds 3 sections, wider than a normal frame but the same dimensions in profile. We usually have a few ready to put into the hives when conditions are right. Dividers are not used with hanging sections.

6.31.3 **Management for cut comb production**. Very little change in management techniques is required to produce cut comb, the important points are detailed below:

- Attention to the foundation is the first item and it is surprising how thick the 'thin' foundation from the commercial suppliers can be. We start looking for ours before Christmas every year and some years we have been very dissatisfied. If it is too thick a compromise is to cut it in half diagonally as a large starter, then at least half has a thin mid-rib but the bees do change from worker to drone at the diagonal. Never use thick foundation [if it is wired it is thick] and pull the wires out before packing into containers.
- Ideally the cut comb needs to be 1½" thick when finished by the bees, this means using a minimum ultimate spacing between frames in the supers of 2" to consistently achieve this.
- Most beekeepers produce a mixture of extracted honey and cut comb. The best way of doing this is to select the best frames for cut comb before extracting the remainder. This does mean using a radial extractor because all super frames will be equipped with thin foundation and no wiring. On average, about 2 or 3 frames per super come up to the standard required.
- For those beekeepers who work the rape, this is the first good flow for getting new foundation pulled out and new comb built. We get all our stocks on the rape to pull out a complete super during this flow, invariably the second super. These frames when extracted and cleaned out by the bees are used for cut comb in the main flow or on the heather later in the year.
- One problem with colonies on small single brood boxes [eg. nationals] is that the brood nest [actual brood] is very close to the top of the frame even with a queen of moderate laying ability. The pollen stores adjacent to this brood extend into the first super above the queen excluder and pollen is found in these frames. Cut comb should be free of pollen; it

has a bitter taste and is quite unacceptable for sale. It is alarming to see how much cut comb is sold with the odd cell of pollen. With single small brood boxes all cut comb is likely to be reaped from the second super and above.

6.32 An account of the preparation of sections and cut comb honey for sale.

6.32.1 **Sections**. The basic requirements to be observed when selling sections are as follows:

- The sections must be fully sealed [capped in wax by the bees]. If open half filled cells are present, the section is not suitable for sale; honey is hygroscopic and these cells will absorb water.
- The woodwork should have been treated with paraffin wax before going into the hive, which will then leave unstained woodwork when the propolis is scraped off.
- Each section should be held up to the light and any with cells of pollen should be rejected.
- All sections should be put into the freezer [minimum -10°C] for a period of 24 to 48 hours to ensure that any Braula coeca and wax moth eggs are killed. If this is not done, it is possible for the section to be ruined while on the shelf awaiting sale.
- Before packing and after cleaning, the section should be weighed and the net weight noted for labelling.
- The packing should be in specially made cardboard boxes with a built in cellophane window on one side to display the sealed comb.
- After packing into the container it must be labelled correctly in accordance with the law. The special cardboard container usually describe the product adequately and the only addition required is the name and address of the producer [telephone number may be included] and the net weight in both ounces and grams.

6.32.2 **Cut comb**. This involves more work for the beekeeper and can be very well produced to very badly produced during the preparation for sale. The basic requirements are as follows:

- The combs of honey from the supers are selected for thickness, good cappings and freedom from pollen contamination. Any combs that do not meet these criteria should be used only for extraction.
- A flat board or 'formica' surface larger than a super frame is required for preparing the cut comb. The tools required are a Price [comb] cutter [size to suit the clear plastic container; nb. two sizes now available] or large sharp knife and a kitchen spatula for cutting the comb and moving it after cutting.
- The comb in the frame is inspected and placed in the middle of the board with the best side upwards. The knife is run around the inside woodwork of the frame cutting the whole comb free from the frame which then simply lifts off. This can be returned to the bees for cleaning up with the rest of the wet supers.
- The comb is then cut into pieces to exactly fit the containers. This is easily done with the Price cutter which not only cuts it to the exact size but provides a tool to place it accurately in the container all in one operation. If a Price cutter is not available the comb can be cut by knife and placed in the container by using the spatula. A fair degree of expertise is required to not only cut it to the right size but then to get it into the container without damage to the comb. Before the cut comb is put into the container, it should be drained for a few minutes to allow the honey to run off the cut edges. Using the Price cutter does not allow this to be

done easily. There should not be a line of honey in the bottom of the container after it is filled with comb.

- Needless to say the surface of the comb should be immaculate and free from drips of honey and any damage.
- Similar to sections the packed cut comb should be frozen for a couple of days before being removed for labelling after any condensation on the outside has evaporated.
- The labelling requirements are the same as for sections but this time the containers do not embody a description of the product [when will the suppliers wake up?]. The minimum description must be 'COMB HONEY' followed by 'net weight' in ozs. and gms. A black on white 'Ablelabel' with a space to put the actual weights in by hand is acceptable. A second label with the producer's name and address is necessary. There must be more illegal labelling in cut comb than in any other honey product on the market.
- There is a certain amount of wastage from one BS shallow frame when used for cut comb; the surplus can usefully be used as chunks for 'chunk honey' and it is advisable to have some 1 lb. jars handy for the chunks, the jars to be filled later with honey when extraction has been completed.

6.32.3 **Other points of interest**:

- To prevent granulation if comb honey has to be stored for long periods, it should be kept in the freezer at low temperatures. Honey granulates fastest at 57°F and above and below this temperature granulation is slower. Higher temperatures are not good because of the aging effects [diastase and HMF] of the honey and so low temperature storage is the best.
- Heather honey is an ideal honey for cut comb because it is thixotropic [in jelly form until stirred when it will become liquid and flow for a short time and reverting later back to jelly] making it easy to cut up with no waste of the actual honey. Draining is unnecessary.

6.33 Suitable methods of decapping super combs.

6.33.1 **General**. There are a variety of methods for decapping combs, some for the hobbyist and others for large scale operation where a permanent installation is necessary. Some of the more common methods are as follows:

- Various types of knives used either hot or cold and with sharpened edges or serrated cutting edges. The common feature is the length of blade which requires to be about double the width of the super frame.
- Heating for the knives can range from the simple method of dipping into hot water to electric elements designed into the blade or steam being passed through the blade.
- The electric carving knife with the two reciprocating blades adjacent to each other works well if cleaned after each cut in hot water.
- Decapping machines based on a heated reciprocating blade in a fixed position, the comb being passed over the blade.
- Flailing machines which rip the capping off are also very efficient for large scale operations.
- There are a variety of multi-pronged forks for scraping off the cappings, [useful on badly drawn comb. All the methods require some form of tray or receptacle to receive the cappings and process them.

6.33.2 **Uncapping trays**. The essential feature of these trays is to provide a receptacle for the cappings to fall into in order to separate the honey from the wax capping. The two basic methods are cold straining through a suitable gauze or melting the cappings with the honey and allowing them to separate on cooling.

• The Pratley tray uses the heating principle. It is constructed in stainless steel with a water bath and heating element on the underside. It is expensive and probably one of the worst designed pieces of equipment available. There is no thermostatic control for the heater so one is continually switching on and off. The tray is built on a slope with the thin part of the wedge [and minimum thermal mass] at the place where maximum heat is required, with the result the whole thing clogs up and slows down the extracting process. The separated honey is heated to such an extent it is only good for cooking.

• The cold straining method seems to be as good as any, the honey is not ruined and the cappings can be washed for mead making or given back to the bees to clean up.

• None of the devices makes provision for locating the frame over the uncapping tray and it is necessary to fix a bar across with a suitable hole about ¼" deep to rest the frame in while actually cutting the cappings off.

• The beginner to beekeeping should be warned of the pitfalls in this area before he parts with his money; unfortunately most of the experienced beekeepers have learnt the hard way.

6.33.3 **The actual operation of removing the cappings**.

• The important thing is to arrange the set up so that the actual operation is undertaken comfortably with no strain, with everything at the right height and everything to hand. It is as well to study the flow of the work from the full supers coming into the extracting room to the empty ones going out.

• Having two people on the extracting makes life much easier, one decapping and the other operating the extracting machine.

• Most of the books recommend that the frame is held at an angle of about 30° to the vertical and the cut made downwards away from the end of the frame which is being held. The other way is to cut upwards [which we find easier]. In both cases the cappings fall away from the frame into the uncapping tray and do not stick to the uncapped comb.

• Again, most of the literature states that the cut should be made just under the capping [in the air space] to minimise the amount of honey removed from the comb. This also is considered to be a matter of preference; we do a straight and level cut across the face of the comb to end up with a set of even combs in the super ready for next year's operations. There are always likely to be some uneven combs and this is a good time to put these to rights. Manley type frames help to provide a straight cut.

• If the knife is heated in hot water it is as well to have a clean damp cloth close by to wipe the knife dry before making a cut, continual drops of water will increase the water content of the honey.

6.33.4 **Other points**.

• Wax cappings should be separated from the honey and rendered down carefully and the wax used for showing. It is unadulterated with other hive products and requires little cleaning.

• The cappings can be cleaned up by the bees in say a Miller type feeder with a grill to allow the bees to access both sides. They may require a stir after a couple of days to let the bees get to the inner sticky portions.

• Extracting is a sticky thankless task and care should be taken to keep everything scrupulously clean and tidy.

• Any honey which has been heated in the Pratley tray should only be used or sold as cooking honey [baker's honey].

6.34 The principles of honey extractors, both tangential and radial.

6.34.1 **General considerations**. The principle involves two forces which occur when an object is rotated [eg. a ball on a string]. The first is the centrifugal force which acts away from the centre of rotation and the second is the centripetal force which acts towards the centre of rotation. In the case of the ball, the centrifugal and the centripetal force act on the ball and because they are equal the ball stays where it is on the end of the string. In the case of a honey extractor, only the centrifugal force acts on the honey and it moves outwards away from the centre, hits the drum wall and trickles down to the bottom under another force, gravity. The centripetal force, of course, is acting in the rotating frame of the extractor and being rigid does not move.

6.34.2 **Tangential extractors**: generally have a cage whereby the frames for extracting can be inserted at a tangent to the circle of motion allowing only the honey on the outside face to be removed during operation of the extractor. The honey on the inner face is pushed on to the comb septum during operation thus necessitating the frame to be reversed in order to complete the extraction. The salient points of this type are as follows:

- With full frames the first side can only be partially extracted at slow speed otherwise the weight of honey on the inner face is likely to break the comb.
- The combs must be reversed to extract the second side and then reversed again to complete the extraction of the first side. This is time consuming.
- It is virtually essential to have wired combs for use with this type of extractor.
- The number of frames that can be extracted is limited by the size of the cage, 6 being the realistic limit.

6.34.3 **Radial extractors**: are all very similar with a rotating framework designed to hold the frames radially with their top bars vertical and parallel to the sides of the drum and their side bars on a radius of the extractor's rotating framework. The salient points are:

- Honey on both sides of the comb is extracted simultaneously and there is no reversing of frames which makes the process that much quicker.
- Combs that are unwired can be easily extracted without damage.
- The design permits a greater number of frames for a given diameter drum compared with a tangential, 10 being a realistic size for a small model [or one super].
- Speed control is not as critical as the tangential but nevertheless it is necessary to start slowly.
- For a given speed and number of rotations, the radial is inferior on the amount of honey remaining in the comb.

6.34.4 **Other points of interest**.

- Materials for construction range from tin plate, galvanised steel and polythene to stainless steel. These days only the two latter materials are used. Stainless steel costs more but is likely to prove the most economical in the long run.
- Both types can be either hand or motor driven. If the electric motor is used a speed control device is necessary to give a range of speed from 0 to about 400 rpm. The problem arises in getting a device to give high torque at low speed when it is starting up from rest. The horse power [hp] required from the motor is quite small, a fraction of a hp being adequate.
- When purchasing a radial type, care should be taken that it will accommodate the frame the beekeeper intends to use. Many [most?] are designed for ⅞" side bars and the full complement of Manley frames cannot be loaded because of their wide side bars.
- Most tangential types can take a reduced number of brood frames which is often useful; only some of the radial types can do this.
- Both types require to be loaded giving the best dynamic balance. No matter how carefully the frames are selected and loaded it will never be perfect and the extractor at top speed wants to move. There are two solutions, one to screw it down to a platform with castors and let it move or two, to bolt it securely to the floor. Our own is fixed in three places with rigging screws and chain from the top rim of the extractor.
- If a fixed installation is preferred then it is desirable to raise it so that a honey bucket can be put straight under the outlet tap without any lifting.

6.35 Methods of straining small quantities of honey and its subsequent storage.

6.35.1 **The objective**: of straining is to remove solid matter down to a particle size determined by the strainer. There are three types of solids, those that sink to the bottom, those that float and those that remain in suspension. The solids can include wax, bees, grubs, propolis, sugar crystals, wood chips and other extraneous matter. Different strainers will remove different solids.

6.35.2 **Types of strainer** include the following methods:

- Single settling tank. The tap is at the bottom and only those solids that float will be strained out including air bubbles.
- Sump tank with baffles. The input is usually at a low level and the outlet at a high level of a tank containing 3 or 4 baffles giving 4 or 5 compartments. The baffle openings are alternately top and bottom with the honey flowing alternately under and then over the baffles. The surface can be skimmed as required to remove floating debris while the dense solids collect in the bottom of the tank. The tank can be double walled to provide a water jacket for heating if required.
- Wire and cloth strainers. These can be in a variety of formats depending on the scale of the operation. Wire strainers should be made of monel metal and cloth is usually silk or nylon. Commercial strainers [eg. O.A.C. honey strainer] incorporate a series of concentric strainers of different mesh size the honey flowing through the coarse mesh first and the finest mesh last [eg. 12 mesh / inch to about 80 mesh / inch. All straining should be below the surface of the honey to prevent bubbles forming.
- The most simple type used by many small time beekeepers is to strain the honey as it goes into the settling tank. It has one major drawback; it produces a large number of bubbles which are formed when the honey drops from the straining cloth into the settling tank.

6.35.3 **Other points in relation to straining** are:

- The higher the temperature the easier it is to strain with a fine mesh. A temperature of 95 to 100°F is considered to be satisfactory.
- If a honey such as rape is to be stored in buckets and bottled later it is better to complete the fine straining before it is stored. Later it only needs warming to a point where it will flow for bottling; this stage is reached well before all the crystals have melted. In this state it would be impossible to pass it through a strainer without bringing it back to the completely liquid state and thereby heating it unnecessarily.
- Long straining cloths can be an advantage; when clogged, pull across the tank to an unused portion.
- Large commercial honey packing organisations use very fine filters by pumping the honey at high temperatures and then cooling it quickly after straining. Such methods are not a practical proposition for small scale operations but the liquid honey has a wine like clarity which is very attractive to the buyer.

6.35.4 **Subsequent storage of strained honey.** Honey in small quantities should be stored in 30lb. polythene [white only for food] buckets, anything bigger is difficult to melt down again for further treatment and bottling. The old method used 28lb. tins with lever lids with attendant problems of the lacquer and plating becoming faulty and the subsequent rusting.

The main points are:

- The buckets require to be as full as possible to minimise the amount of air trapped in the top.
- Before the lid is snapped shut, the centre of the lid should be depressed onto the honey again to minimise the air content.
- Store at a temperature of 57°F for rapid granulation and then at as low a temperature as possible after it has set. Don't open to check the granulation, do it by feel, the sides of the buckets are quite flexible when the honey is in the run state but very solid when granulated.
- An old refrigerator makes an ideal warming cabinet for reheating when fitted with a fish tank thermostat, a small fan and a couple of electric light bulbs. The fan is necessary to prevent hot and cold spots forming. Our's doubles up as an incubator for queen cells in the summer by resetting the thermostat.

6.36 Methods of small scale bottling and preparation of honey for sale.

6.36.1 **The types of honey**: which are likely to be bottled on a small scale are as follows:

- Run or liquid honey. Because all honey is a supersaturated solution of sugars it will granulate in a greater or lesser time depending on the types of sugars it contains. It is therefore essential that it is heat treated before sale to delay granulation for 6 to 12 months [130°F for about 45 minutes for an average honey]. Partially granulated run honey looks terrible and is a very common fault with the hobbyist beekeeper.
- Granulated honey. This should be of a fine texture and an average honey should be seeded with 10% rape or similar honey which granulates with a fine grain.
- Creamed honey. Again a fine texture is necessary.
- Chunk honey. The comb must be surrounded with a heat treated liquid honey.

6.36.2 Faults that are to be avoided. There is much local honey on sale which should really be taken off the shelf and returned to the beekeeper as being unfit for sale. Many of these jars are labelled with a county association label and thereby brings other beekeepers a bad reputation. We believe that a county organisation should take some responsibility for quality if their label is to be used. The common faults are:

- Partial granulation of run honey.
- Frosting of granulated honey. This a particularly difficult fault to avoid with honey that granulates rapidly - it is caused by shrinkage, generally at the neck of the jar, as the honey granulates and is associated with low temperatures. Honey that granulates slowly or granulates at higher temperatures [greater than 57°F] seldom exhibit this fault.
- Fermentation. A vertical streak in granulated honey with the surface texture rough. In run honey recent bubbles on the surface can be seen. At an advanced stage it has a characteristic unpleasant smell.
- Scratched lids and unclean wads.
- Under weight.
- Incorrect labelling.

6.36.3 Methods to be employed in order to produce a quality product:

• The first requirement is that all honey must be ripe when extracted otherwise fermentation can occur. Heating the honey to a temperature of 130°F for a short time will kill the yeasts but will not reduce the water content sufficiently to make it completely safe from re- infection by other wild yeasts.

• Run or liquid honey:

- Usually bottled and heat treated after it is in the jar by putting the jars complete with wads and lids in a water bath for about 45 minutes at 130°F. The same effect can be obtained by putting the jars in a microwave oven covered with 'clingfilm' in lieu of the metal lids. It will be necessary to experiment to get the power and time right to obtain the correct temperature.

- Clean jars are essential, any minute particles of dust or other solids provide a nucleus for granulation to commence.
- A few jars should be filled after taking the weight of the jars and then re-weighed. Such a test should be carried out at random during bottling. If no light can be seen through the jar, under the lid, when it has been filled it will almost certainly pass the weight test.

• Set or granulated honey:

- Granulated honey should have a fine texture; coarse crystals give the honey a gritty characteristic which is generally not liked by the consumer. Therefore most main crop honey should be seeded with a rape or clover to obtain the fine grain.
- The seed should be melted to a point where it just flows and has a granular texture. 10% should be stirred into the liquid honey, without causing bubbles, and then bottled as for run honey.
- To ensure granulation takes place as quickly as possible the jars should be stored at a temperature of 57°F or as near as possible. Fine granulation minimises the water content between the crystals and thereby minimises fermentation.

- Creamed honey:

    - Is produced in exactly the same way as finely granulated honey except that after it has fully granulated and hard a further treatment is given.
    - If the set honey is heated to 80 to 90°F for approx. 24 hours it will become creamy and smooth. It is then stirred with a special mixing tool avoiding the introduction of air bubbles and then allowed to re-set. After this treatment it will not revert to the hard state.

- Chunk honey:

    - Requires one piece of sealed comb approx. 1"×1"× the height of the jar and filled with heat treated liquid honey.
    - The major fault is to provide too small a piece of comb. The liquid honey needs to be well filtered and 'bright' to show off the chunk in the jar.

- Equipment required:

    - Warming cabinet. 30lb buckets take about 5 days at 90°F to get to a state where the contents will flow [eg. rape]. For other honey to get it to the stage where it is completely liquified ready for straining and bottling about 3 days at 120°F is required.
    - A good set of scales is required and the calibration should be checked regularly.
    - Settling tank to hold about 112 lb together with honey stirrer for stirring seeded honey.

6.36.4 **Other points.**

- When bottling has been completed the jars should be free from stickiness on the outsides.
- Suitable labels should be attached on each jar; note many labels are too big for ½lb jars.
- In the future EEC regulations may require all beekeepers producing, processing and selling honey to be registered. Most beekeepers undertake the processing in the family kitchen. It prompts the question whether it is suitable? Are there two sinks and are the walls and floors washable surfaces? It is likely to be taboo if there is a washing machine installed where soiled linen is washed despite the fact that all the family food is prepared every day in the same room. Beekeepers must be vigilant of current legislation, it is all too easy to transgress unwittingly.

6.37 Details of the legal requirements for labelling and sales of home produced honey.

6.37.1 **Labelling.** For full details see appendix 1; with the advent of the EEC, all regulations are subject to change and all beekeepers should acquaint themselves with the latest regulations and remain on the right side of the law. The important points for labels are:

- The name of the product to be displayed [eg.comb honey, chunk honey, etc.].
- There must be no misrepresentation in words or pictures.
- The producer's name and address must be displayed [telephone number optional].
- The net weight to be displayed both in imperial and metric units.
- The size of the lettering for weights is critical.

Labelling is very much a personal choice, we believe in simplicity and use Ablelabel [32mm ×

63mm] black lettering on gold for all our honey. Once you have a label, 'stick' with it if your honey sells well, the public quickly learn to recognise it as a sign of a good product.

6.37.2 **Sales**. Sales of home produce may take place at the door providing the goods for sale have been produced on the premises. It is therefore necessary for the beekeeper to have his hives in the garden or grounds of his home otherwise he will be contravening the retail licensing laws. If all your honey is sold to a retailer then the apiaries may be sited anywhere. The labelling requirements are identical in both cases.

6.38 Methods of storing comb with particular reference to prevention of wax moth damage and sterilisation against nosema.

6.38.1 **Types of comb to be stored**: are supers and brood comb.

• Supers. These can be stored either wet [with honey] or dry after being cleaned up by the bees and removed from the hive again. This comb consists only of wax and honey [if stored wet].

• Brood comb. This type of comb is very different containing wax, pollen, larval skins and faeces, propolis, etc. making them much more attractive to attack by other insects and mammals.

6.38.2 **The main causes of damaged comb are by**:

• Mammals such as mice, rats, squirrels, etc. which are easily excluded with travelling screens or queen excluders at the top and bottom of the stacks of boxes of frames.

• Insects, the main cause of damage being the wax moths. There are two:

  • The lesser wax moth [Achroia grisella] and
  • The greater wax moth [Galleria mellonella] which is regarded as the major pest. However both can cause very extensive damage in a short time if precautions are not taken.

6.38.3 **Methods of protection against wax moth damage**. There are four methods namely paradichlorobenzene [PDB], acetic acid [80%], heating and cooling. Each of these either kill [k] or have no effect [ne] on the various stages of development ie. egg, larva, chrysallis and adult. Reference to published literature reveals the following:

	PDB	ACETIC	FREEZING	HEATING
Eggs	ne	k	k	k
Larva	k	ne	k	k
Chrysallis	?	?	k	k
Adult	k	k	k	k

T [FREEZING] = 0°C to -17°C for a few hours to a few days depending on temperature and bulk of frames.

T [HEATING] = 120°F [49°C]. Note the melting point of wax ≈ 145°F.

? = no reference could be found in the standard literature.

It will be clear from the above table that the best method is freezing before storage at normal temperatures. No airing of the combs is necessary. A good method for supers where the risk of disease is very low compared with brood frames.

Heating would also be possible but wax is very malleable at 120°F and the temperature control would have to be precise.

Gamma radiation is known to kill all stages in the life cycle but is expensive and not really a practical method for the beekeeper. For the average beekeeper fumigation is the more usual method and it will be clear that both acetic acid and PDB is necessary and is the accepted method of dealing with brood frames.

Brood frames. These should first be fumigated with 80% acetic acid as follows:

- Brood box with frames placed on suitable board [acetic acid attacks concrete].
- Metal runners in brood box well greased to protect from acid fumes. Any metal ends are removed from the frames.
- An empty eke is placed over the brood box and covered by a suitable board.
- The acetic acid is poured onto an old piece of rag in a shallow dish standing on the tops of the frames. The rag extends over the edge of the dish acting as a wick. The fumes are heavier than air and fall through the frames. The amount of acetic acid required is 100 mL per BS brood box.
- All the joints should be sealed with tape to make the set up as air tight as possible.
- Leave for one week and then the frames can be stacked for winter storage.

After fumigation with acetic acid they should be stacked as follows:

- Mouse excluder with newspaper over.
- Sprinkle one dessertspoonful of PDB crystals onto newspaper and place brood box over.
- Cover with newspaper, PDB another brood box, etc. finishing with a screen and crown board.

Supers. It is unnecessary to fumigate with acetic acid and they can be stacked straight away with PDB and newspaper. At the end of the season it is important to get all frames cleaned up, fumigated and stacked for winter as soon as possible. While the weather is warm the wax moth can do considerable damage.

Note that if supers are stored wet it is necessary to to make them beeproof if they are stored outside otherwise they must be in a beeproof shed or room. When the frames are wet, they are not attractive to wax moth so they can be stored without PDB. They do become very damp [honey hygroscopic] and tend to grow mould during the winter.

6.38.4 Other relevant points:

- PDB crystals should never come into contact with the wax comb. They will dissolve in it and it is impossible to remove it from the wax.
- Sterilisation with acetic acid is the approved method of cleaning comb infected with Nosema spores, so the storage treatment ensures that there is no risk of the spread of infection the next season when the comb is reused. Other pathogens are killed with acetic acid such as chalk brood fungus spores so it is good beekeeping practice to make this storage the norm.

- Rather than have the trouble of stacking and sealing boxes for acetic acid treatment it is probably better to have a permanent installation to hold as many frames as required. It can be custom built and made completely air tight thereby using less acetic acid. The sterilisation box automatically can become a frame storage box in the winter.
- After any fumigation, combs should be well aired before re-use in the hive.

6.39 Wax moth damage to stored comb.

6.39.1 **Greater wax moth [Galleria mellonella].**

• The adults have a wing span of 1" to 1½" and enter the hives at night to lay eggs. The eggs hatch to larvae and when fully grown they are about ⅞" long and quite distinctive with a dark head. The larvae pupate, usually in a boat shaped groove chewed into the woodwork of a frame. The chrysallis eventually hatching to the adult form. The damage is caused during the development of the larval form. The larva has the ability to digest wax but it also needs protein which is obtained from pollen and larval debris of the honeybee.

• The life cycle is approximately egg - 7 days, larva - 15 days, pupa - 28 to 32 days. The times are very variable and depend on temperature [egg to adult on average is 50 to 54 days].

• In warm weather there is the possibility of all the comb in a full brood chamber being turned to dust in about 14 days and much of the woodwork damaged if it is off the hive with no bees. A strong colony will not tolerate the moth and keeps itself in a healthy state and no damage is caused. Weak colonies can be damaged with the bees in occupation.

• These moths are generally only troublesome in UK when the comb is not in use, they can however be very real pests in tropical climates.

• It is possible to introduce bacteriological control by impregnating the wax foundation with spores of Bacillus thuringiensis which kills the wax moth larvae. This form of control has been used in USA but is not practised widely in UK at the present time [It is sold under the trade name CERTAN].

6.39.2 **Lesser wax moth [Achroia grisella].**

• This moth which is much smaller and has a wing span of about ¾" to 1" and the larvae do not cause damage to the woodwork. The larvae still consume and digest the wax comb and while doing so they produce a large web of silk tunnels.

• It is not so much of a pest as the greater wax moth but can completely ruin comb if an infestation occurs and no protective measures are taken.

6.39.3 **Death's head hawk moth [Acherontia atropos].**

• This is a magnificent looking moth with a skull and cross bones marking on its thorax on the dorsal side. It is attracted to bee hives and is quite rare in UK; it originates from N. Africa and Spain. It is worth having a look at one in a natural history museum.

6.40 Small scale methods of recovering beeswax from both comb and cappings.

6.40.1 **Types of wax collected**: in the season include cappings during extraction, brace and burr comb collected during inspection and manipulations and finally old combs which are being discarded.

• The cappings are of high quality and need very little treatment to clean them up for use. They should be used solely for show wax, making cosmetics [eg. cold cream] and for high quality blocks for sale.

• Quite a considerable amount of brace and burr comb is collected during colony manipulations and should be lumped in with the old combs for rendering. Since these two sources contain wax contaminated with propolis [which cannot be removed] it is only suitable after home rendering and cleaning for making foundation, candles, etc. There is no way of cleaning wax at home comparable with large scale commercial operations with heated pumps and filters.

6.40.2 **Processing cappings**.

• The cappings will be initially separated in a decapping tray or similar device with a mesh basket to allow the honey to drain off.

• The cappings can be given back to the bees to clean up or washed to make mead with the washings.

• Finally the cappings are melted and filtered through lint as a final cleaning process. The wax should not be heated above 150°F to prevent discolouration. To save the natural colour of the wax, metal containers should be avoided.

6.40.3 **Processing old comb**.

• After the wax has melted it adheres to the old larval skins and except by pressing or centrifuging at high temperatures some wax will inevitably be lost. These two methods are generally unsuitable for home operation.

• There are two suitable methods for home use, these are the solar wax extractor and the steam boiler.

• The steam wax extractor is a boiler with a mesh cage suspended inside with a drain at the bottom for the wax to run off. The device has a water reservoir which is converted to steam which melts the comb and wax inside. It can be driven by gas or electricity.

• The solar wax extractor is probably the best device, it costs nothing to run and can be made for a few £s. They are expensive to buy from the equipment suppliers. Most of the doubling glazing merchants have lots of old panels available which they are only too glad to sell and produce a better job than the commercial models available.

• A second melting in a saucepan of soft water leaving it to cool and float on the water. Any dross ['slumgum'] can be scraped off the bottom and it is ready for use. Further cleaning by filter if required.

6.40.4 **The solar wax extractor.** As this method is by far the most suitable and there is little meaningful information in the general literature, a few points of interest on the device are:

- When it is normal [at right angles] to the sun on a cloudless day the energy collected is approx. 1kW per metre square [compare with a 1kW electric fire].
- The correct angle to the horizontal [α] is given by the formula: α = Lat - Dec where Lat = latitude and Dec = sun's declination. The sun's declination varies from 0° in March to +23° in June to 0° again in September. It goes on to -23° in December [the winter solstice] and then to 0° again the following March. The average value for the summer months March to September ≈ 18°. On the South Coast of UK with a latitude of say 50°, then α = 32°.
- Black bodies absorb the most radiation so the inside should be black or as dark as possible, and not white, for the greatest efficiency.
- The glazing should be double and preferably a sealed unit and very good insulation [fibreglass for lofts] is also necessary for efficiency.
- Take care not to burn yourself on the metal inside, the temperature is well over the boiling point of water on a good day.
- Because the temperatures are so high it acts as a steriliser and will kill most pathogens. It is therefore a good idea to make the extractor large enough to take one or two whole brood frames complete with comb. To ensure that foundation wax is free from AFB and EFB pathogens, the wax should be maintained at 100°C for 30 to 60 minutes and this should be done as a separate operation outside the solar wax extractor if there is any doubt, particularly if colonies have been treated with antibiotics for EFB.
- Keep the glass clean for maximum efficiency. The inner surface is the most difficult to clean as it develops a thin film of vaporised wax and propolis; methylated spirit will clean it off.

6.40.5 **Other points of interest.**

- One very old type of extractor [the M & G extractor] is still to be found and works well with care and controlled heating. It is a metal drum and old comb is put inside and filled up with water and a filter is tied across the open top. Around the outside and as part of the device there is a large rim to catch the contents as they are pushed through the filter. The rim has a spouted outlet for the wax. The wax is forced out by hydraulic pressure when water is poured into the high spout, the bottom end of which is connected into the lower part of the drum. If care is not used it is claimed that the ceiling is likely to receive a wax treatment.
- Physical properties of beeswax are:

    - Specific gravity = 0·95.
    - Melting point = 147·9 ± 1°F.
    - Solidifying point = 146·3 ± 0·9°F.
    - Insoluble in water.
    - Soluble in chloroform, ether, benzene, etc.
    - When stored cold for some time it develops a surface bloom which is not a mould or mildew - its cause is not well understood.

- Most appliance suppliers will purchase rendered beeswax or exchange it for equipment.

6.41 A method of preparing home made foundation.

6.41.1 **General**.

• There are two methods suitable for making foundation at home. These are by 'press' or by 'Herring' methods [both are described below].

• Foundation is required in two thicknesses, one for wiring [thick] and the other for cut comb [thin] which requires to be very refined as it will eventually be consumed. We do not believe that it is possible to make thin foundation of sufficient quality at home for two reasons:

  • without recourse to a mill with rollers the thickness is too variable and more often than not out of limits.
  • without special pumps and filters it is not possible to clean the wax sufficiently unless, of course, only capping wax is being used.

• It is not easy to measure the thickness of foundation and a good guide is to check the number of sheets to the lb. BS brood frames for wiring have about 8 sheets to the lb. and Commercial about 5 sheets.

• Historical - Mehring is credited with making the first wax foundation in Germany in 1857 not long after the discovery of bee space. Weed, in the USA, made the first automatic mill for producing foundation on a large scale.

6.41.2 **The foundation press method**.

• The equipment consists of a bottom tray with a tinned copper platen soldered inside. The lid, with a similar matching platen, fits inside the tray and is hinged to one of the longer sides. They are expensive to purchase.

• In use, molten wax is poured into the tray and over the lower platen and the lid quickly closed. As soon as the lid is closed the excess molten wax is poured off. When cooled [a matter of seconds] the moulded foundation is removed from the press; this should remove easily if the dies have been treated before moulding with a suitable release agent [washing up liquid or similar].

• The sheet of foundation needs to be trimmed accurately to size; this is usually done with a trimming board, which is the exact size required, and a sharp knife or roller cutter.

• The principle is simple but there is a fair degree of skill required in getting the right working temperature for the wax, the right amount of detergent in the water for the release agent and the speed of working with the right amount of wax being poured into the press.

6.41.3 **The Herring method**.

• This method, due to Mr. Herring, is probably the best for use at home. It has not been written up in the general literature and Herring himself has not been given sufficient credit for producing, at low cost [about 10% of the cost of a press], a simple and efficient device.

• The Herring die consists of two flexible moulds, made of rigid polythene, and joined on one of the shorter sides, the whole being an inch or two larger than the foundation size being produced.

• To produce a sheet of foundation, a flat sheet of wax is placed in the die and the whole rolled through the mangle of a washing machine [albeit an old one].

• There are two methods of producing the flat sheet of wax, either by:

   • dipping a board once or twice in a container of molten wax [to control the thickness] which produces two sheets one on each side of the board,
   • or by pouring onto a flat board with edging all round except for a small opening to run off the excess wax.

In both cases the boards are well soaked in water to allow the wax sheet to come away from the board.

• The wax sheet needs to be warm [about 95°F] and malleable to be rolled and impressed. The die needs to be soaked in a release agent as for the press method.

• A trimming board is used to finally trim the sheets to size, again as for the press method.

• Other points on the method;

   • Old wringers are becoming hard to come by these days.
   • We have found a thermostatically controlled reservoir, for warming the die and the wax sheets for pressing, makes the operation easier. The reservoir is filled with water and the release agent [a few drops of washing up liquid].
   • It takes us about half a day to make approximately 100 sheets of foundation [15 to 20 lb. of wax] using the pouring method.
   • After the sheets are dry they are stored flat between pieces of newspaper.
   • Take care to position wax sheet in the die squarely otherwise the final foundation will be incorrectly orientated in the frame.

6.41.4 **Wiring** - see section 6.5 for full details.

6.42 The possible effect of stings and be able to recommend first aid treatment.

6.42.1 **General**.

• It is the mature worker bee, 14+ days old, which is capable of injecting bee venom into its chosen victim. Queens only use their sting on other queens and drones have no stinging apparatus. It is a means of defence used by the honeybee, generally as a last resort. After stinging mammals, the honeybee leaves behind the stinging apparatus, including the 7th abdominal ganglion, thus terminating its own life.

   • Stings can be minimised as follows:

   • Beekeepers maintaining stocks of docile bees by culling the queens producing nervous bees and those with strong defensive traits eg. followers.
   • Handling colonies correctly i.e. no jarring of frames, no fast movements, no squashing of bees or banging of hives. Bees are very sensitive to vibration.

- Only open stocks under good weather conditions. Thunderstorms or approaching rain clouds definitely affect the temper of the bee.
- Always wear a veil to protect the eyes, the nose, the mouth and ears where there is a proliferation of mast cells [mast cells are pharmacological 'time bombs'; when they rupture they release powerful chemicals that can effect various tissues nearby, such as blood vessels or smooth muscle].
- Use protective clothing of correct material eg. cotton. It is said that woollen garments or garments dyed blue are best avoided. In our opinion the temperament of the bee is the predominant factor and not the clothing.
- Dispersing the sting pheromone by the application of the smoker to the site of injection does discourage other guard bees being attracted to the site of the first sting. For stings on the hands, place a hot part of the smoker on the sting area to evaporate, as quickly as possible, the volatile pheromone and then smoke the area to mask any remaining smell.
- Refrain from using perfumes, aftershave lotions, hair shampoos, nail varnish, hair sprays and other similar cosmetics prior to working with bees as these have been known to elicit a stinging reaction from bees due to the similarity in the chemical makeup to the isopentylacetate in the alarm pheromone of the sting chamber.
- Site stocks as far away as possible from the general public.

6.42.2 **The effect of stings**.

• Most beekeepers at the first sting may experience pain, reddening of the skin and swelling. The severity seems to vary depending on the site and the number of stings. But over the years a natural immunity will be built up and the discomfort and swelling will be minimal.

• Extensive swelling may occur taking 12 hours to reach its maximum and 2 or 3 days to resolve. These symptoms may indicate an increasing sensitivity to bee venom.

• A generalised reaction with symptoms of difficulty in breathing, skin rash, palpitations, vomiting and faintness occurring within minutes of a sting indicates a severe reaction [anaphylaxis] requiring emergency medical attention.

6.42.3 **First aid treatment**.

• The barb should be removed as soon as possible by a scraping action of a knife, hive tool or finger nail, not by squeezing between the fingers. This will only inject more venom [See notes 2.3.5.].

• Away from the stocks of bees, apply ice to the site of the sting especially if it is situated where there is little spare skin for expansion eg. ear, nose or tip of finger, will bring some relief of pain.

• Application of calamine lotion, steriod creams or crushed leaves of the mallow plant [marsh mallow (althaea officinalis) has long been used to relieve inflammation] may give some relief. Antihistamine creams are best avoided as their repeated application can cause severe skin sensitization.

• Aspirin tablets may reduce pain and inflammation. Piriton tablets contain antihistamine and may lessen the symptoms. If in doubt consult a doctor.

• To the non-beekeeper a sympathetic attitude will often solve the immediate problem but should there be any severe reaction as may be the result of a sting close to or on the eye ball, in or around the mouth or neck then it is always safest to obtain medical advice.

6.42.4 **Other points of interest.**

• Most beekeepers expect to receive the odd sting and take no further action beyond removing the barb.

• About 15% of the population have an atopic constitution and roughly 50% of the children of two atopic parents will be similarly afflicted. This group includes individuals with a personal history of hay fever, eczema, asthma, allergic rhinitis and urticaria. They may show progressive worsening in their reactions to stings developing general symptoms such as nausea, skin rashes and respiratory difficulties. Medical aid should be sought immediately should anyone show these symptoms.

• Taking non inflammatory drugs eg. aspirin or piriton under medical advice one or two hours before working in the apiary does reduce the reaction to stings. It should be remembered that antihistamines cause drowsiness so that driving the car is unwise after this kind of medication.

• For the hypersensitive person who wishes to continue beekeeping or for the members of the beekeeper's family who are exposed to the danger of being stung and are hypersensitive a course of immunotherapy can be arranged through their family doctor.

** ** ** **

APPENDIX 1.

SUMMARY OF THE LAWS APPLYING TO THE SALE AND SUPPLY OF HONEY PRODUCED IN THE UNITED KINGDOM.

The following Acts and Regulations apply; they should be referred to if an exact wording is required:

- Food and Drugs Act 1955
- The Labelling of Food Regulations [1970 as amended]
- The Honey Regulations 1976
- The Materials and Articles in Contact with Food Regulations 1978
- Weights and Measures Acts 1963 to 1979
- The Weights and Measures [Marking of Goods and Abbreviations of Units Regulations 1975 as amended]
- The Weights and Measures Act 1963 [Honey] Order 1976
- Trade Descriptions Acts 1968 & 1972
- The Trade Descriptions [Indication of Origin] [Exemption No. 1] Directions 1972
- Consumer Safety Act 1978 Glazed Ceramic Ware [Safety] Regulations 1975

LEGAL DEFINITIONS.

"Honey" means the fluid, viscous or crystallised food which is produced by honeybees from the nectar of blossoms, or from secretions of, or found on, living parts of plants other than blossoms, which honeybees collect, transform, complete with substances of their own and store and leave to mature in honeycombs.
"Comb Honey" means honey stored by honeybees in the cells of freshly built broodless combs and intended to be sold in sealed whole combs or in parts of such combs.
"Chunk Honey" means honey which contains at least one piece of comb honey.
"Blossom Honey" means honey produced wholly or mainly from the nectar of blossoms.
"Honeydew honey" means honey, the colour of which is light brown, greenish brown, black or any intermediate colour, produced wholly or mainly from secretions of or found on living parts of plants other than blossoms.
"Drained honey" means honey obtained by draining uncapped broodless honeycombs.
"Extracted honey" means honey obtained by centrifuging uncapped broodless honeycombs.
"Pressed honey" means honey obtained by pressing broodless honeycombs with or without the application of moderate heat.

METHODS OF SALE.

When sold by retail not prepacked, eg. from a bulk container, honey should be sold by net weight.
When prepacked ready for retail sale in a quantity of more then ½ oz., the net weight of honey in the container should be one of the following: 1 oz, 2 oz, 4 oz, 8 oz, 12 oz, 1 lb, 1½ lb or a multiple of 1 lb.
Chunk honey and comb honey may be packed in any quantity.

MARKINGS ON CONTAINERS.

Honey should be prepacked for retail sale or otherwise made up in a container for sale only if the container is marked with the following information:

- An indication of quantity by net weight in both imperial and metric units.
- The name or trade name and address of the producer, packer or seller.
- A description in one of the following forms:

    a) Honey   b) Comb Honey   c) Chunk Honey
    d) Baker's Honey or Industrial Honey.
    e) The word "honey" with a regional, topographical or territorial reference eg.
    Devon honey, Honey from South Devon, Moorland Honey.
    f) The word "honey" with a reference to the blossom or plant origin e.g. Heather
    Honey, Lime Honey.
    g) The word "honey" with any other true description e.g. Honeydew, Pressed Honey,
    Set Honey.

In wholesale transactions of containers of a net weight of 10 kgs, or more, a separate document showing the required information is sufficient if it accompanies the container.

METHODS OF MARKING CONTAINERS.

- The imperial and metric indications of quantity should be of equal size. The minimum height of any figure used is:

    For a quantity of 1 oz - 2 millimetres.
    For a quantity of 2 oz or 4 oz - 3 millimetres.
    For a quantity of 8 oz, 12 oz, 1 lb, 1½ lb or 2 lb - 4 millimetres.

The units of weight used should be at least half these heights.

- The two quantity indications should be distinct but in close proximity, the imperial quantity being shown first. Nothing should be inserted between them.

- The permissible units of weight with their permitted "abbreviations" are:
    pound - lb; ounce - oz; kilogramme - kg; gramme - g.
No other abbreviation should be used. In the case of imperial units, the letter 's' may be added to indicate the plural.

- There should be one type space between the figure and the unit used.

- All required markings should be clear, legible, conspicuous and indelible.

CONTAINERS.

Containers should be made of materials which under normal and foreseeable conditions of use do not transfer their constituents to the honey in quantities which could endanger human health or bring about a deterioration in its aroma, taste, texture or colour or bring about an acceptable change in its nature, substance or quality. This applies to containers which are in contact with the honey and to the containers which are likely at some later time to be in contact with the honey.

Certain ceramic materials may present particular risks. Packers are asked to obtain an assurance from their suppliers that containers comply with 'The Materials and Articles in Contact with Food Regulations 1978' and 'The Glazed Ceramic Ware (Safety) Regulations 1978' and 'The Glazed Ceramic Ware (Safety) Regulations 1975', if applicable.

MISDESCRIPTION.

There are two basic types of illegal misdescription; the direct and the indirect or misleading. The direct misdescription should be obvious and can be fraudulent. A simple example would be to describe Australian honey as "Devon honey". Careful thought will avoid indirect misdescription. Examples of such misdescription could be:

> a) An illustration of bees collecting nectar in a moorland setting on honey which is not from moorland.
> b) The statement "Produced in Devon" applied to honey which is blended in Devon from honeys of various origins which may or may not include Devon.

The following guideline should be followed.

• Any reference, direct or indirect, in words or by means of any pictorial device to the blossom or plant origin should only be applied to honey derived wholly or mainly from the blossom or plant indicated.

• Any such reference to the regional, topographical or territorial origin of the honey should only be applied to honey which originated wholly in the region, place or territory indicated.

• Description which have legal definitions should be applied only to products which fall within the generally accepted meanings of those description references.

• Descriptions and other references which have no legal definitions should be applied only to products which fall within the accepted meanings of those descriptions or references.

• A honey may fall within more than one definition, in which case it may be described as being any one or more. For example, a pressed Devon heather honey may be described as "Honey" or "Heather honey" or "Pressed heather honey" or "Devon blossom honey" or any other true combination of words.

• Honey produced outside the UK which has a UK name or mark should be accompanied by a conspicuous indication of the country in which the honey was produced. Blends of honeys from two or more countries, which may include the UK, may be accompanied instead by a conspicuous indication that it was produced in more than one country.

COMPOSITION OF HONEY.

• There should be no addition of substances other than honey.

• The honey should as far as practicable, be free from mould, insect debris, brood and any other organic or inorganic substance foreign to the composition of honey. Honey with these defects should not be used as an ingredient of any other food.

• The acidity should not be artificially changed and there is a legal maximum level of acidity.

• Any honeydew honey or blend of any honeydew honey with blossom honey should have an apparent reducing sugar [invert sugar] content of not less than 60% and an apparent sucrose content of not more than 10%. Other honeys should have an apparent reducing sugar content of not less than 65% and an apparent sucrose content of not more than 5%.

• Honey with a moisture content of more than 25% should not be supplied.

• The maximum water insoluble solid content is:

      for pressed honey: 0·5%
      for other honey: 0·1%

• The maximum ash content is:

      for honeydew honey and blends containing honeydew honey: 1·0%
      for other honey: 0·6%

BAKER'S OR INDUSTRIAL HONEY.

Honeys of the following descriptions should be labelled or documented only as "baker's honey" or "industrial honey":

• Heather honey or clover honey with a moisture content of more than 23%.

• Other honey with a moisture content of more than 21%.

• Honey with any foreign taste or odour.

• Honey which has begun to ferment or effervesce.

• Honey which has been heated to such an extent that its natural enzymes
have been destroyed or made inactive.

• Honey with a diastase activity of less than 4, or, if it has a naturally
low enzyme content, less than 3.

• Honey with an hydroxymethylfurfuraldehyde [HMF] content of more than 8mg. per kg.

IMPORTANT NOTE.

Government regulations are constantly changing and being updated. It is therefore important that beekeepers make themselves familiar with up to date information before processing, packing and selling honey for retail sale. With the run up to 1992 in the EEC further changes may be envisaged particularly in the area of metrication.

** ** ** **

APPENDIX 2.

WAGTAIL DANCES & DIRECTION.
[see section 1.10]

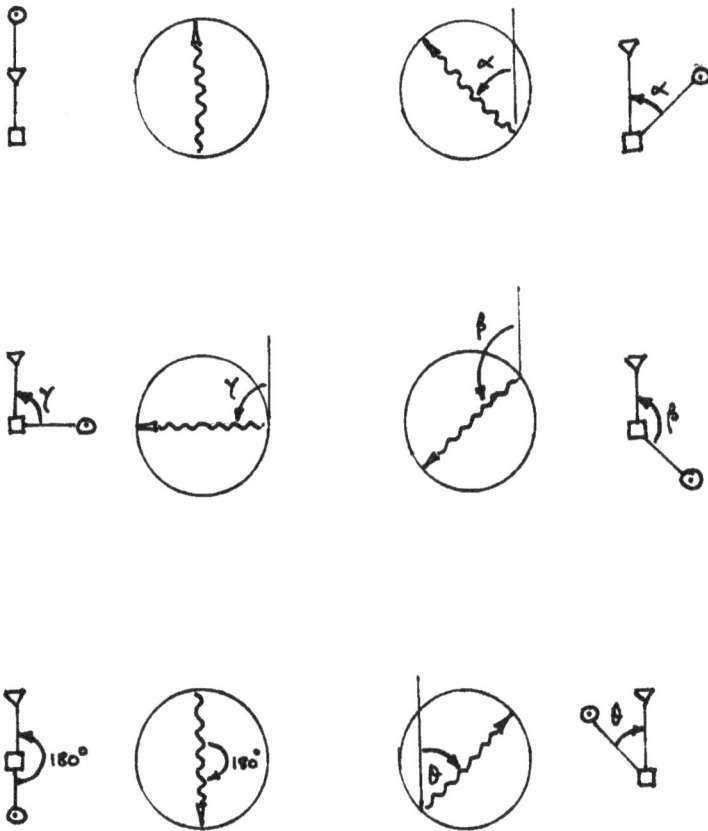

Legend: ⊙ - Sun   ☐ - Hive   ▽ - Source

Notes:

    • When the food source is to the left of the sun [viewed from the hive], the bees dance on the comb to the left of the vertical.

    • Similarly, if the food source is to the right of the sun, the bees dance to the right of the vertical.

    • If the food source is in the same direction as the sun, the bees dance upwards on the comb.

    • If the food source is in the opposite direction to the sun, the bees dance downwards on the comb.

AVERAGE COLONY POPULATION CYCLE.
[see section 1.11]

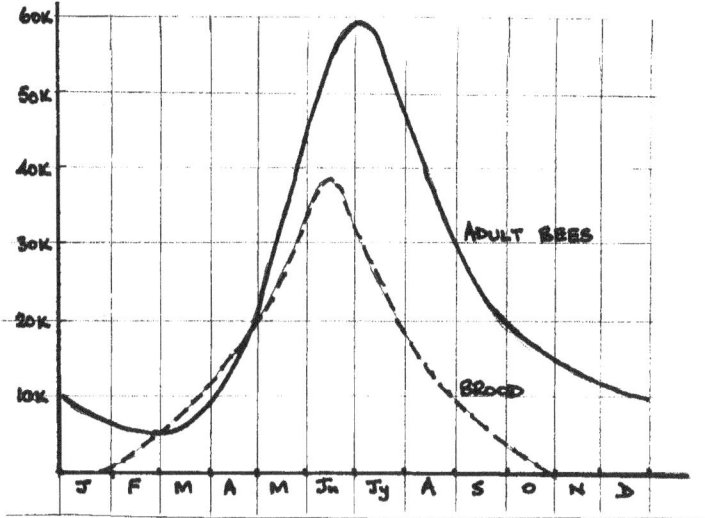

Notes:
- Brood = adult bees twice per year.
- Brood > adult bees from Feb. to April. This is a very critical time in the annual colony cycle [nb. the danger of chilling brood and not having enough adult bees to incubate this brood].
- Brood peaks in early/mid June.
- Adult bees peak end June/start July (3 weeks after brood peak); this is the time the main flow usually starts when the maximum foraging force is required.
- After the main flow the population starts to decrease, rapidly at first (old foragers dying off) then more slowly as the winter bees (6 month life) start to appear in the colony.
- The minimum adult population occurs c. end February (c. 5000).
- The maximum population will vary from 40,000 to 60,000 depending on the fecundity and strain of the queen.
- The population builds up on the "spring flow" often using all the income and storing very little.
- The maximum adult population stores very large amounts in a short time for winter (much less brood to care for).
- The reduced population allows adequate reserves for winter.
- Brood rearing ceases in late autumn and starts again after the winter solstice when the days start to lengthen.
- There is a continual decrease in population throughout the winter so dying bees are not abnormal at this time. The healthy colony removes any that die in the hive.

It should be noted that the graph is a representation of average conditions and the local flora and climatic conditions will modify it accordingly. Similarly, these local variations mean a peaky graph and not a smooth curve as shown.

** ** ** **

ANATOMY & OTHER DIAGRAMS.

EXOSKELETON

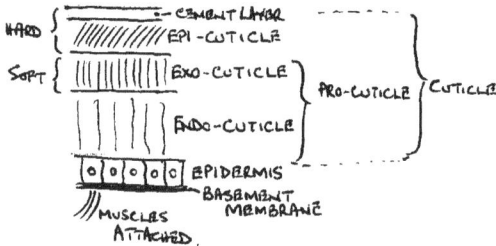

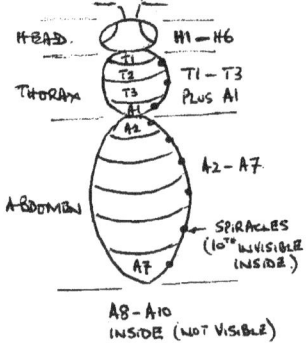

MOUTHPARTS

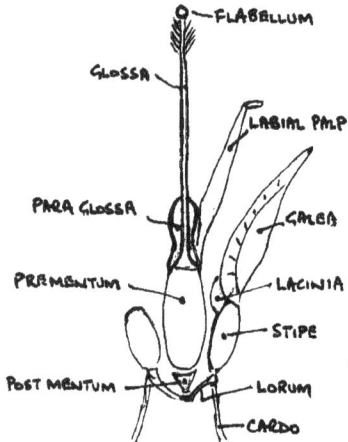

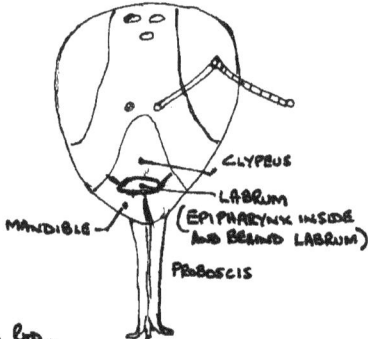

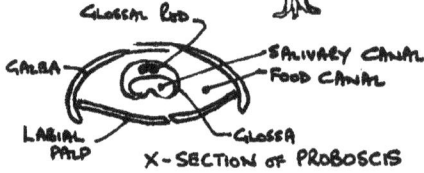

X-SECTION of PROBOSCIS

ANTENNA

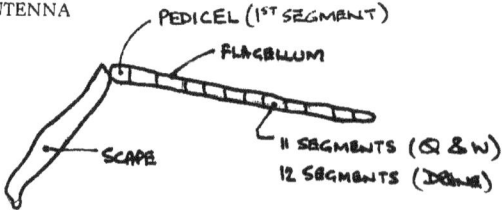

WINGS

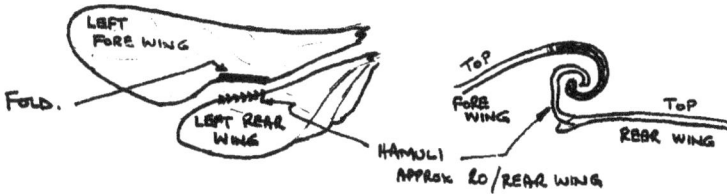

LEGS

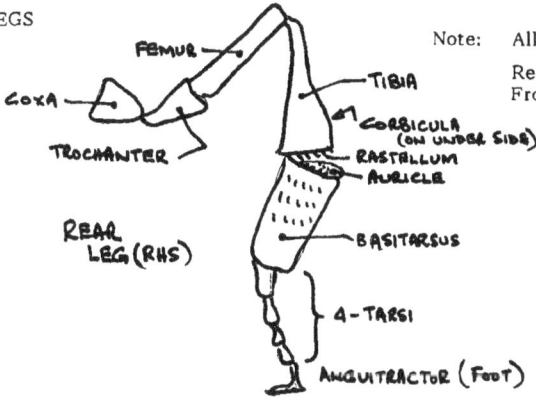

Note: All legs similar

Rear – pollen press
Front – antenna cleaner

STING

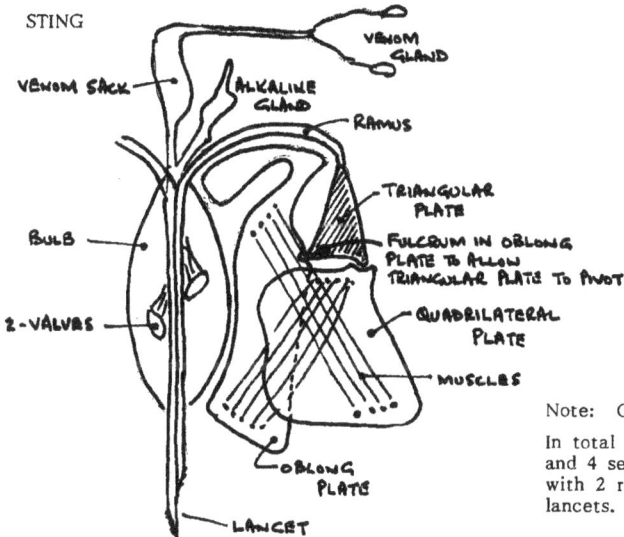

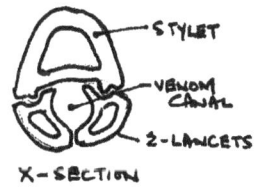

X-SECTION

Note: Only one side shown

In total there are 6 plates and 4 sets of muscles with 2 rami to drive the lancets.

ALIMENTARY CANAL

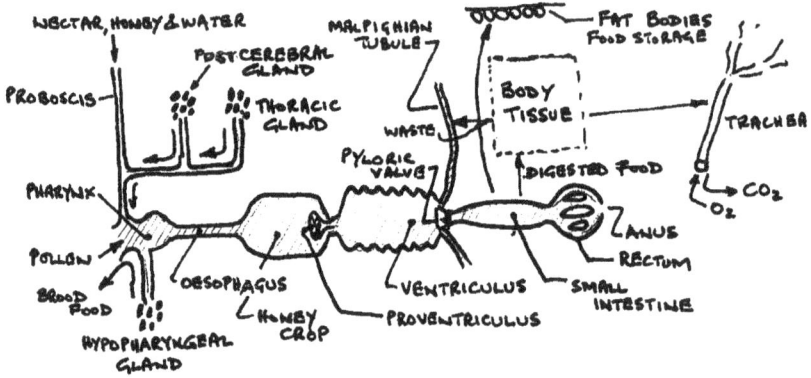

CIRCULATORY SYSTEM

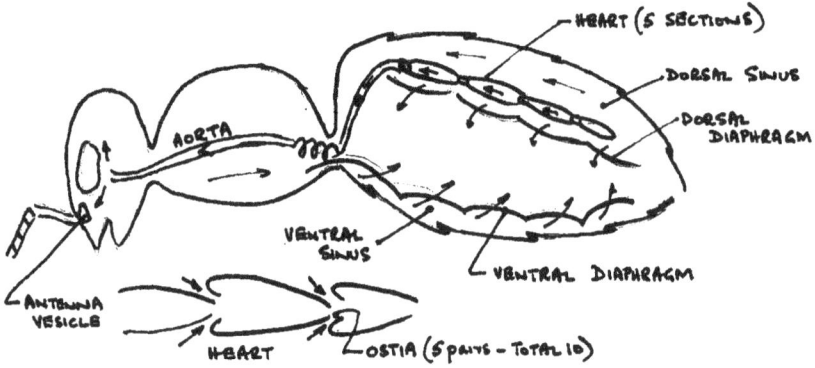

RESPIRATORY SYSTEM

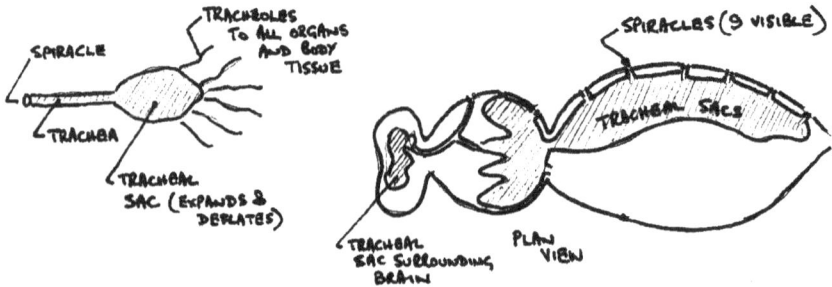

WAX GLANDS

WAX GLAND
A4
WAX POCKET
STERNITE
A5
WAX SCALE
A6
MIRROR
A7

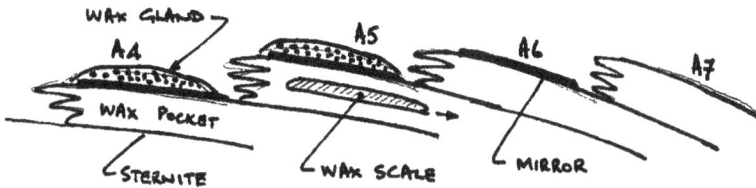

4 PAIRS OF WAX GLANDS (STERNITES A4-A7)

HONEY USAGE

STORED HONEY
80 : 20

HONEY SAC

NECTAR
30 : 70

50 : 50

METABOLISED IN THE COLONY

RATIO VERY VARIABLE

WATER
0 : 100

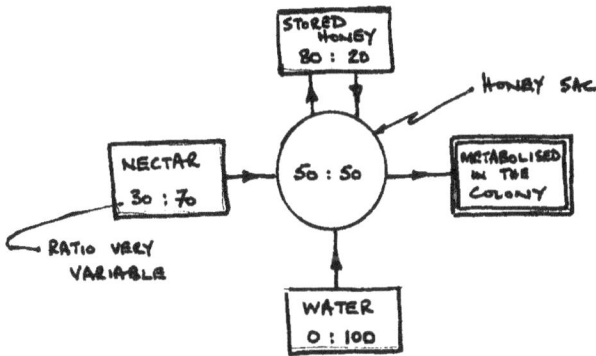

RATIOS SHOWN ARE
SUGAR : WATER

FLOWERING PLANT

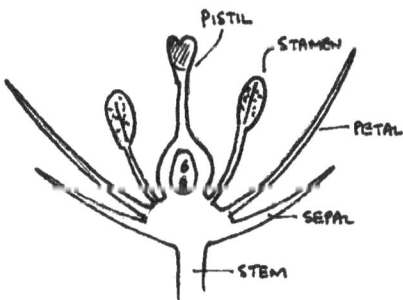

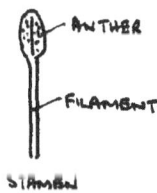

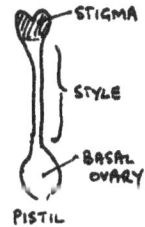

PISTIL
STAMEN
PETAL
SEPAL
STEM

ANTHER
FILAMENT
STAMEN

STIGMA
STYLE
BASAL OVARY
PISTIL

APPENDIX 5

COLONY INSPECTIONS (TIMING)
[See section 6.16.4]

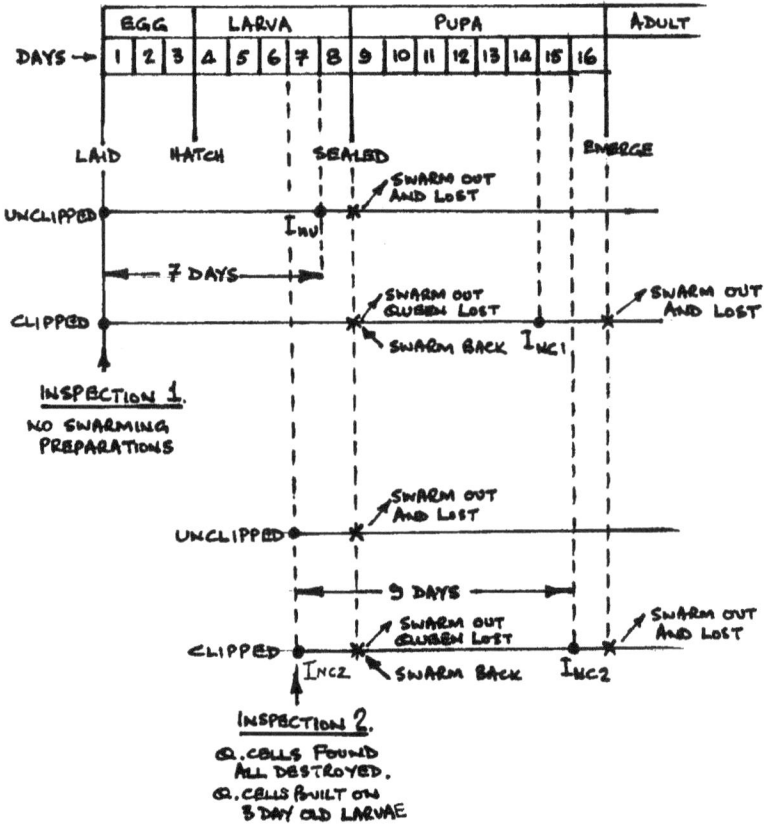

SWARM CONTROL INSPECTIONS:

Unclipped queen   Every 7 days

Clipped queen   Every 14 days – when last inspection showed no
                                    swarm preparations.
                     Every 9/10 days – after queen cells destroyed.

RECOMMENDED BOOKS

Books are expensive to buy and one of the objectives of these study notes has been to minimise this aspect. There are three books which we consider essential for the intermediate student which should be purchased from the outset; these are:

Anatomy & Dissection of the Honeybee	**IBRA**
The Hive & the Honeybee	Dadant & Sons
Guide to Bees & Honey	**Northern Bee Books**

Other books which we consider essential reading for the intermediate examination are as follows:

The Social Organisation of the Honeybee	**Northern Bee Books**
The World of the Honeybee	**Collins (also available from Libraries and Second Hand dealers**
Some Important Operations in Bee Management	**Northern Bee Books**
Beekeeping at Buckfast Abbey	**Northern Bee Books**
Swarming - its Control & Prevention	**Northern Bee Books**

The BBKA publish a book list for those studying for the examinations and it is worth consulting. It includes quite a few books which are out of print, difficult to find and expensive to buy [eg. Behaviour and social life of honeybees by C.R.Ribbands].

The recommendation of books is a very subjective matter and the student will have to make up his own mind in the long run. It is a good idea to have a quick glance through any book on bees that you can lay your hands on and then you can decide. One book that we read years ago was the Beekeeper's Folly by John R.Ratcliff; we enjoyed it and it is on our list to buy when we find a copy going cheap!

Information leaflets can also be classed under books and all the up to date leaflets from MAFF should be obtained. Some of those out of print are useful and informative if they can be found. There is a good selection of BBKA leaflets and it is possible to obtain a catalogue of these in order to choose which to buy [they are of nominal cost]. Unlike many of the out of date books, most of these leaflets contain reliable information.

** ** ** **

APPENDIX 7

TYPICAL EXAMINATION QUESTIONS.

The examination comprises two written papers each of 2½ hours as follows:

Paper A - Practical [dealing with the more practical aspects of beekeeping].
Paper B - Scientific [dealing mainly with the theoretical and scientific aspects including anatomy].

The papers follow a set pattern as stated below:

- Five questions have to be answered [½ hour per question]; the first one is always on disease and is compulsory.
- Any other 4 questions may then be answered from the total of 7 questions on each paper.
- There are 20 marks for each question with a break down of these marks beside each part of the question. This is important information for the candidate; it tells him roughly where the effort has to be applied.
- Only pen and pencil are allowed to be used, the pen having black or blue/black ink. The reason for this is so that the papers may be photocopied for marking without losing any of the information which may happen with some colours and some machines.
- Generally there is a question in each paper writing brief notes on 5 topics taken from a list of between 7 and 10. In our opinion these questions can be very deceptive, take care before committing yourself [eg. a few notes on a hive tool compared with the exoskeleton both of which appeared in March 1988 papers - it would be possible to write a book on the exoskeleton and you have no idea which points the examiner will place emphasis on].

Past papers can be purchased from the BBKA and we recommend that both papers for the last 5 years are purchased and the questions attempted at home; it is good practice and you will be better prepared for the examination.

We append below typical questions from both types of papers.

PRACTICAL.

1   [a] List the honeybee larval diseases and their causative agents. [7]
        [b] Describe the signs of two of them. [7]
        [c] Describe the treatment for the two diseases chosen. [6]

2.   An apiary of 6 stocks is to be requeened during the flying season with queens produced in the same apiary, describe in detail:
        [a] the method of queen rearing to be used and the equipment required, setting out the time schedule for the operation, [12]
        [b] the method to be used to requeen the colonies. [8]

3.   Write brief notes on 5 of the following: laying workers, propolis, prevention of damage to comb by wax moth, robbing, supersedure, damage to colonies by mice and how to exclude them, drone laying queen. [4 for each]

4.   Describe the steps that can be taken to minimise bee losses when prior notification is

received of the use of toxic chemicals on flowering crops near an apiary. Assume the colonies cannot be moved. [20]

5. [a] Describe in detail a method of making nuclei - (1) to receive mated queens, and (2) to receive sealed queen cells. [12]
[b] List the various uses of nuclei. [8]

6. Describe:
[a] the possible causes of robbing, including those actions by the beekeeper that may cause the outbreak, [7]
[b] the problems that an outbreak of robbing might cause, [7]
[c] methods that can be taken to terminate an outbreak. [6]

7. Give an account of:
[a] the processing of extracted honey for retail sale in glass jars in liquid and set form, [12]
[b] the legal requirements for labelling. [8]

SCIENTIFIC.

1. [a] Name the mite that which can affect honeybee colonies in UK. [3]
[b] Describe the signs which may indicate the presence of the mite and how this can be confirmed. [6]
[c] Describe fully the action to be taken if infestation is confirmed and the steps to minimise the spread to other colonies. [11]

2. Describe with the aid of diagrams the structure and segmentation of the exoskeleton of the honeybee. [20]

3. Write short notes on 5 of the following with diagrams where applicable: spiracles, parthenogenesis, queen substance, hamuli, lancets, nasonov gland, malpighian tubules.
[4 for each]

4. [a] With the aid of diagrams, name and describe the position and function of the following: exocrine glands in the worker honeybee, hypopharyngeal, salivary, mandibular. [15]
[b] Describe the major differences between nectar and honey. [5]

5. [a] Describe with the aid of simple diagrams the mouthparts of the worker honeybee. [10]
[b] Explain exactly how the different parts function and the uses to which they are put during the life of the honeybee, both inside and outside the hive. [10]

6. [a] Draw a sketch of a flower visited by the honeybee and name the parts. [10]
[b] Describe the role of the honeybee in the process of pollinating a flower. [10]

7. [a] Give an account of the likely sequence and timing of events in an undisturbed colony which swarms during a period of fine weather in early June. Your account should cover the period from the commencement of swarming preparations to the emergence of the offspring of the new queen. [12]
[b] List the factors which might have caused the colony to swarm. [8]

** ** ** **

APPENDIX 8.

MANIPULATING A COLONY OF HONEYBEES.

This appendix is to assist those who are attempting the Basic Examination and deals with that part of the Basic Syllabus seldom addressed in other text books. It is the way we handle our own bees. As you become more experienced you will probably develop your own techniques.

1. The need for care when opening and handling a colony.

Honeybees are primitive insects and they are much the same as they were 10 to 20 million years ago, wild and untamed. You will probably have seen in your reading reference to the domestic bee as opposed to the bumble bee; the honeybee has not been domesticated and is unlikely to be. Man has only learnt, mainly through trial and error, how to manipulate honeybees for his own purposes.

1.1 Great care must be exercised when opening a colony, the need for this care is as follows:

- To ensure your own safety.
- To ensure the safety of others [neighbours and passers by].
- To ensure the safety of pets, any other domestic animals and any other livestock.

1.2 The temperament of honeybees can range from being very docile to very aggressive. Understanding the reasons for this wide range of reaction to intrusion into their nest is helpful in determining our approach:

- The honeybee's temperament is part of its genetic make up and in no way can it be changed except, of course, by breeding.
- Good tempered bees can become aggressive very quickly [during a manipulation].
- Bees are more aggressive during bad weather, eg. rain and electric storms [thunder and lightning]. Bees are known to be irritable if stocks are sited under power lines so they probably have the ability to measure electrostatic charge in some way. The orientation to magnetic fields is known to have an effect on comb building.
- Queenless colonies are also more aggressive than queenright colonies.
- Bad handling can, and certainly does, make bees bad tempered.

1.3 Bad handling techniques can be corrected by the beekeeper. The points to avoid are as follows:

- Any vibrations, bangs knocks,etc.
- Any fast, quick or speedy movements.

Initially, when learning to handle bees, go very very slowly making all actions deliberate and purposeful.

1.4 It is necessary to be able to detect any signs of restlessness or change in the colony's behaviour during a manipulation. Points to watch for are as follows:
• Activity at the entrance is the first indicator. Study this before touching or smoking the colony so that you have a mental picture of the norm.
- Watch the entrance from time to time while you are manipulating the colony; this must become second nature to you.
- Any bees starting to collect at the entrance or starting to collect up front of the hive is a

warning sign of trouble.
- There are indications inside the hive also.
- While in the brood chamber, rows of eyes between the frames spells trouble; rows of 'bottoms' means a contented colony
- When bees start to dart off the frames, while they are being removed or replaced, is another warning that they are getting 'fed up' with the performance.

2. Reaction of smoke on a colony. Providing the smoke is cool and has a pleasant wood smell (a somewhat subjective statement!) they immediately and instinctively gorge themselves with honey. With full honey sacs they are in a docile state. the origin for this obscure, but the popular story is absconding of the colony in the wild state at the onset of a forest fire.

2.1 There are many acceptable fuels that can be used in the smoker and every beekeeper seems to have his favourite:

- Wood shaving from a plane are ideal for starting the smoker; they can be lit directly, like paper, with a match.
- Softwood planer chippings from a sawmill or timber yard are good fuel.
- Dry rotted wood is good and can be broken up easily by hand ; it possibly burns a bit on the hot side.
- Dry grass cuttings are an old favourite.
- Sacking or hessian burns steadily with few sparks.
- Corrugated paper rolled into a cartridge used to be popular but most of the packing material is now impregnated with chemicals to prevent it burning.

2.2 Some thoughts on the smoker [the beekeeper's most important tool]:

- Purchase the largest you can afford,
- Preferably in copper or stainless steel.
- Have spare fuel and lighter to hand during a colony manipulation.
- Before starting a manipulation ensure the smoker is fully fuelled.
- There is only one place for the smoker during a manipulation and that is between your KNEES; make this practice second nature to you when manipulating , the smoker will then always be in the right place and ready for use quickly.
- Keep it in your possession throughout the manipulation.
- Extinguish the smoker when you have finished (ie. plug the nozzle with green grass). Beware starting fires by careless practices.

3. Personal equipment required to open a colony. The following equipment is required by every beekeeper:
- A good quality veil, hive tool and smoker are essential.
- The following are optional:
    - Gloves, bee suit, wellies, etc.
    - A couple of cover clothes can be very useful.

Because bees are often handled very badly and because insufficient effort is made to cull for bad temper and rear docile bees, beekeepers protect themselves thoroughly and often are responsible for innocent bystanders getting stung. **PLEASE PONDER THIS POINT.** I remember, on one occasion, parking our car at the side of the road near a hedge and, on opening the door and getting out, I was stung twice on the face in as many seconds. Further investigation revealed a beekeeper in his apiary on the other side of the hedge and out of sight from the road. I shudder to think what could have happened to an allergic child. The beekeeper, of course, was dressed up to the nines.

4. Keeping a colony under control. A colony can only remain under the control of the beekeeper if:

- The bees are in the hive.
- The colony has been subdued with smoke.

4.1 When opening a colony for say a routine inspection the following sequence should be followed:

- Approach the colony quietly and gently place any spare equipment on the ground near the hive.
- Observe the normal activity at the entrance.
- Smoke at the entrance [6 good puffs of the smoker]. Don't be timid with the smoker at this stage.
- Wait for at least two minutes [you should know why].
- Very gently remove the roof and place it on the ground, upside down, behind the hive with a corner pointing to the back of the hive. This may sound a bit pernickety but it is now all ready in the right position to put the supers on.
- Now study the record card, check that you have all the equipment required and that you are sure what has to be done.
- Two small puffs of smoke at the entrance; we're coming in now!
- Ease the supers up on one side with the hive tool, tilt and smoke well between the supers and the queen excluder.
- Lift off supers and place gently on the upturned roof. No bees will be squashed as there are only 4 small points of contact.
- Smoke the bees down so that the QEx is clear of bees and then gently lever and twist it off. Hold it over the hive and shake any remaining bees off and place in front of the hive.
- NOW PUT THE SMOKER BETWEEN YOUR KNEES! ready for use.
- Remove the dummy board, shake off any bees into the top of the hive with a sharp tap and lean it up at the rear of the brood chamber.
- A small puff of smoke to clear the bees at the lugs of the first frame [where you want to hold them] and gently remove it, inspect and replace hard up against the side of the brood chamber.
- Don't forget to glance at the entrance.
- If the bees are tending to well up and fly off, smoke them down.
- Repeat the operation with every frame always keeping the bees down in the hive with the smoker. No jerking or bumping or fast movements are acceptable.
- Replace the dummy at the opposite end of the brood chamber and lever the frames and dummy with the hive tool so that there are no gaps between the spacing devices. This will prevent a build up of propolis which would change the frame spacing.
- Smoke the tops of the frames clear of bees and replace the QEx.
- Replace the supers carefully without squashing any bees.

Note: if there are no supers on the colony, check that the queen is not on the underside of the crown board when it is removed.

4.2 If the bees are difficult to keep down in the hive with smoke then use cover cloths, but avoid them if possible. Sunlight/daylight also tends to subdue bees. If cover cloths are needed most of the time then, in our opinion, the strain of bee could usefully be changed.

4.3 THERE IS AN ACQUIRED ART IN THE USE OF SMOKE: WATCH CAREFULLY A COMPETENT BEEKEEPER AT WORK, THE ART IS EASY TO LEARN.

4.4 If you feel the colony is getting out of control, then close it down before you get into a tail spin. It is the sensible thing to do and is good beekeeping practice.

4.5 Do not wear gloves to manipulate your colonies, have them by you and on standby for an emergency. If you are frightened of a sting or you are allergic to stings, then beekeeping is not your forté.

5. **Take a sample of worker bees in a matchbox or similar container**. It is quite remarkable the number of candidates who come forward for the basic examination and they have never taken a sample and often have no idea how to proceed.

5.1. Collecting a sample for diagnosis with a matchbox is easy. Have it open in one hand, holding a frame of bees in the other and slowly manoeuvre the open box over the bees and wriggle it slowly shut. If there are plenty of bees on the frame, it will just hold the required 30 bees for the sample. Old bees are required for disease diagnosis so always take the bees from the end frames of the brood chamber.

** ** ** **

APPENDIX 9.

CONSUMPTION OF STORES DURING WINTER.

Many years ago we weighed the colonies in our apiaries throughout the year. This was done for a number of years and the results were very informative, for example, in the summer it showed that the main flow consistently started on virtually the same date every year. In winter it showed how the consumption of stores was directly related to the weather. Shown below is a graph of the stores consumed from October to March in a typical year and is typical of all the measurements made on our colonies. The results are actual and were obtained by measurements made with a steelyard.

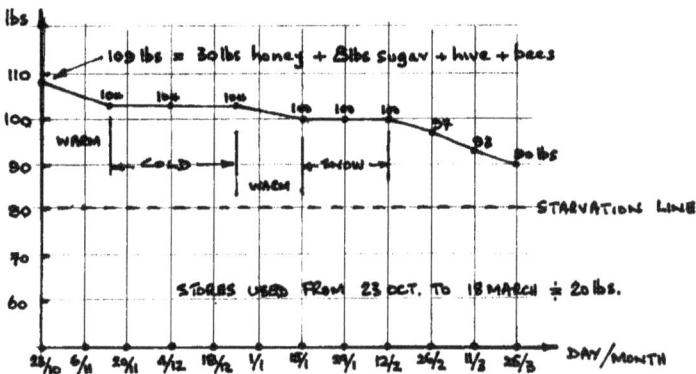

Some points of interest from the graph:

• This particular colony overwintered on a BS brood box and one super with no queen excluder.
• The colony was fed 8lb. of sugar which is equivalent to 10lb. of honey, ∴ total stores = 40lb. in October.
• The starvation line has been drawn in to indicate when the stores in the colony have reached 10lb. We regard this as the critical point and if there is no income, the colony may require to be fed.
• The autumn was warm up to mid November and c. 6lb of stores were used.
• Then the weather was cold [40°F and below] up to Christmas and hardly any stores were used [ounces not lbs]. Many beekeepers find this surprising.
• Christmas to mid January the weather was warmer [45°F+] and bees were flying on cleansing flights. Stores are being used.
• Come February when it gets warmer and brood rearing is increasing, stores are starting to be used very rapidly.
• From the beginning of March the beekeeper must be alert to possible stores shortage in his colonies. If the colony has 40lb of stores in October we have never experienced the necessity to feed in the spring.
• The graph demonstrates the nonsense of feeding candy on Christmas Day.
• Further, removing the roof alerts the colony, puts up the temperature of the cluster, shortens the life of the bees and induces them to use more stores.

** ** ** **

— 172 —

BAILEY FRAME CHANGE

This is a method of transferring a stock of bees on to sterilised comb or foundation. It is recommended for use in the spring or early summer for a stock of bees which has been infected with Nosema apis, Malpighamoeba mellificae or Ascosphaera apis [chalk brood]. The spores of these diseases remain dormant on the comb and will be re-cycled by the house bees cleaning the cells of the brood nest, infecting the larvae in the case of chalk brood or the adult bees during the exchange of food in the case of nosema or malpighamoeba.

During a period of fine spring weather when the bees are foraging daily and the brood nest is expanding, [a weather pattern which we are likely to experience during April/May] the colony should be re-arranged on the original site as follows:

- The queen plus the brood frame and bees on which she is found is placed in the middle of a clean brood box complete with frames of comb or foundation. Mark this frame with a drawing pin.
- The gap in the frames of the bottom brood box is closed and the dummy board moved up.
- The queen excluder is placed on top of the old brood box.
- A "U" shaped spacer 18"x18"x ⅞" thick [the same as the entrance block] is placed on top of the queen excluder thereby providing a space for a new entrance in the same direction.
- The clean brood chamber is placed on top of this spacer with a clean crown board, feeder and the old roof.
- Now close the original entrance and reduce the new entrance.
- Feed the stock the same evening with a gallon of syrup [eg. strength 1 Kg sugar to 1 litre water] into which is mixed 166mg Fumidil "B" [for nosema infected colonies].
- Keep an accurate record card and destroy any queen cells if they appear in the bottom brood box after 7 days.
- After 3 weeks all the brood should have emerged from the bottom brood box. It can be removed together with the queen excluder, "U" spacer, old floor board and entrance block.
- A clean floor board and entrance block should be given to the new brood box.
- The marked comb in the top box should be moved to the side of the brood box for removal when empty of brood.

Other points to note:

- After the first re-arrangement the bees will orientate to the new entrance by exposing their Nasonov glands and fanning vigorously. This may attract bees from other hives in the apiary. During such a procedure on a fine sunny day in 1985 our re-arranged colony attracted two overwintered nuclei complete with queens which were promptly killed on entry!
- Splitting the brood nest does put the bees under stress especially if it is a small colony and it may cause chilled or chalk brood.
- If foundation is used for the new brood box it may be necessary to continue feeding. Care must be taken to avoid robbing [see 6.12 & 6.26].
- Always check that the Fumidil "B" is not time expired and has been stored in the dark.
- Follow the maker's instructions when mixing the Fumidil "B", its activity is lost if the syrup is hotter than 120°F.

- The queen is likely to be infected with nosema and the colony should be re-queened as soon as possible after the change of combs has been completed.
- A sample of bees should be taken in the autumn before the winter feeding to monitor the success of procedure.
- All the old combs will need to be sterilised before re-using, the other contaminated hive parts should be scraped clean and flamed with a blow lamp.

** ** ** **

www.ingramcontent.com/pod-product-compliance
Lightning Source LLC
Chambersburg PA
CBHW081400270326
41930CB00015B/3362